Simple 8th Grade Science Investigations

Solutions for doing science in the classroom.

Christopher P. Garside

Seven Sides Publishing

Seven Sides Publishing has a mission to improve teaching and the understanding of science. To contact us, send an email to simpleinvestigations@sevensidespublishing.com or visit our website at sevensidespublishing.com.

ISBN: 9798406753903

Published by: Seven Sides Publishing, Cypress, TX.

Table of Contents

Introduction

To help teachers teach science through investigations, Seven Sides Publishing has provided a series of lab manuals for Elementary Science, Middle School Science, Physics, Chemistry, Biology, Environmental Systems, and Earth & Space Science. These manuals are a rich resource for structure and investigations. There is a shortage of user-friendly labs that easily allow teachers and students to perform experiments quickly. Too many labs have too much busy writing within them, where teachers do not want to take the time to read everything to figure out if it would be good to use with their students. If teachers do not want to read it, do you think the students do? So we have taken a lot of the traditional labs that have been around for decades and simplified them; so they are easy to read and perform. We have also added some new original labs that have never been seen before. There have been efforts to try to have teachers do more investigations with their students, but there is no plan or solution to deal with the real issues teachers have in preparing to do this. The book How to Teach Science Through Investigations has the plan, and the Simple Investigations Lab manuals have the solutions so students can learn science through investigations with minimal effort. Teaching science through investigations will make your classrooms more efficient, where students learn content and practice skills simultaneously. Science is a process of doing. Doing this process is the most efficient way for students to learn science and be able to use it in the future. We live in a culture where science-literate people are needed for jobs, but too few can be found. If you incorporate these investigations with virtual investigations (that we will point you to in each section of this manual), skill/math practice, and concept maps, you will not need to fill in gaps by giving lectures. All content can be learned through investigations and practice. Remember, we only remember 5-20% of what we hear. That 20% is when you are really interested in the content. But hearing practices no science process skills and does not activate any higher cognitive thought. Lecturing is not a good option. We remember 75-80% of what we do/experience and 90-95% of what we teach. Investigations allow us to keep our students in these higher retention percentages. Teaching through investigations also works because students spend more time in class at higher Bloom's Taxonomy levels, staying in zones C and D on the Rigor Relevance Chart when they perform investigations. And if you add the physical way they are stimulated with the hands-on experience, you cannot deny the level of learning will be much higher when students perform investigations. This manual gives you the resources you need to teach 8th Grade Science through investigations.

We separated each of these sections in this manual as you may divide your class units. We will follow the 8th Grade Science TEKS to make it easy for you to find the labs you want and need for your classes. We include concept maps at the front of each section that shows the vocabulary and visual clues to how concepts relate to each other; this is a great way to organize information. It talks to the students to see how ideas work together, making it easy to chunk information to use at higher cognitive levels. At the beginning of each lab, we put the materials you will need in boldface in the directions; this saves time for your lab preparation. There is also a safety question in boldface just after that for you and your students to evaluate. It says, "Looking at the material and lab we will be using, what are the safety precautions we should take to protect ourselves and materials during this investigation." Make sure to read the lab so you can appropriately help answer this question with your students.

Virtual Labs

Hands-on labs are not the only way for students to learn science, but they are the most effective. However, virtual labs can be used with these hands-on labs. Many investigations physically cannot be done hands-on, so some experiments will have to be done virtually. There are three sources that I have used in the past that have a good number of resources. **Physicsclassroom.com** and **PhET.colorado.edu** are free to everyone and are great to use. **Physicsclassroom.com** has teacher notes and activities/exercises that guide students through Physics and Chemistry Interactives. You can find them under the simulation and open, download, or print the PDF. They also have a series of Concept Builders that are a tremendous virtual practice that can replace those worksheets that help students practice concepts, math, and skills. They can be hard to find, so above the list provided is the section where they can be found (underlined and in italics) on the website. **PhET.colorado.edu** has a variety of activities of different levels that you can explore to go through their simulations. They are also easy to download and print. **ExploreLearning.com** is expensive, but the quality of its product is much higher than the other two. When you click on a Gizmo, you can also click on lessons and find the Student Explorations that go with each Gizmo that you can modify, download, and print. They are written at a very high quality, making the students think like a scientist. At the end of each section of this lab manual, we include a list of virtual labs from these organizations that would be great to use with our investigations. Please remember virtual labs should never replace hands-on labs. If the students can learn the content live, that should be the priority because it is more of an experience that will be remembered. There are many other virtual simulations out there, but none so far have moved me to use them over the three I have mentioned here.

TIPPERs

TIPPERs are great for students to explore and think about different scenarios for each concept of Physics and Chemistry. These help students think outside the box, apply concepts to real life, and think about how multiple concepts would be used together. I suggest you get the books of TIPPERs to practice and discuss after completing these labs and investigations.

Probe-ware

This manual has lots of labs that use probe-ware. Students must learn how to use probe-ware; this means teachers need to know how to use probe-ware. Many companies use digital probe-ware with all the research, development, testing, and forensic testing they do; this increases potential career opportunities that help students become more marketable for jobs when they are familiar with using probe-ware. Hooking everything up is just as easy as charging your phone. When I was a High School Science Technology Coach and researched which companies and devices would be the most user-friendly to students, I found using Vernier Probe-ware was better for high school students, but PASCO seemed better for middle school students. Both are giants in the probe-ware industry for education. Since I am more familiar with Vernier, I will be referring to Vernier Probe-ware. However, PASCO would be a great alternative.

Interfaces are devices that the probes are connected to that talk with the program (Logger Pro) that displays the data. I found the most economical and friendliest way for students to see the data from probe-ware is to use the Vernier LabQuest Mini interface hooked up to a computer with Logger Pro. LabQuest Mini has multiple ports that are needed in many labs. They are the least expensive, so they are better on the budget. They require no batteries, so they are easy to transport if you need or want to. The other interfaces are more expensive, require batteries if you are going outside, or the stand-alone devices have a smaller screen to see the data, with less flexibility to manipulate the parameters like changing the time of data collection or changing units if you want to change or modify an experiment. Some costly wireless probes and interfaces may be easier to use if you do not mind the cost. A computer screen is much bigger and makes it easier to see the data, so this is my preferred setup. But using any interfaces will work fine for these labs.

Connecting the Probe-ware

To hook them up, you will plug your probe into one of the channels or the sonic on the interface. If the plug does not fit in smoothly, either you are plugging it upside-down or trying the wrong port. Take the little chord that looks like it would go into your phone and plug that into your interface. Take the other end, and plug it into a USB port on your computer. Open up Logger Pro on your computer. If everything is hooked up properly and the computer and interface are working properly, you will see a green button at the top of the computer screen that says "Collect." Many of the labs have preset settings in Logger Pro. You will use the manila folder at the top left of the toolbar in Logger Pro to find the folders and files you will be instructed to go to for these specific settings for different labs. Whenever you get the physical equipment, they will have detailed instructions in the box they come in on how to hook them up if you are still confused. They will also have instructions on how to calibrate the probes if needed. There are a few probes that require frequent calibration. If we use any, it will be discussed in the lab directions. The more you use probe-ware, the easier it gets to set up. I usually only have to show my students twice to have them be able to set the equipment up on their own. But as you are showing them, have them physically do it. You can also find detailed instructions online at Vernier.com. Many more detailed labs can also be found there under lab ideas.

You also can use standard equipment like spring scales for force sensors or thermometers for temperature probes. Because schools want to integrate more technology, we wrote these labs to use probe-ware wherever applicable. Because they are so simple, these labs can be modified to fit whatever equipment you have. There are very few labs that I have used in my career that I did not alter how I presented them. One reason we wrote these labs this way was to customize them to the Texas TEKS and National Standards. We wrote them the way we thought teachers would want to use them.

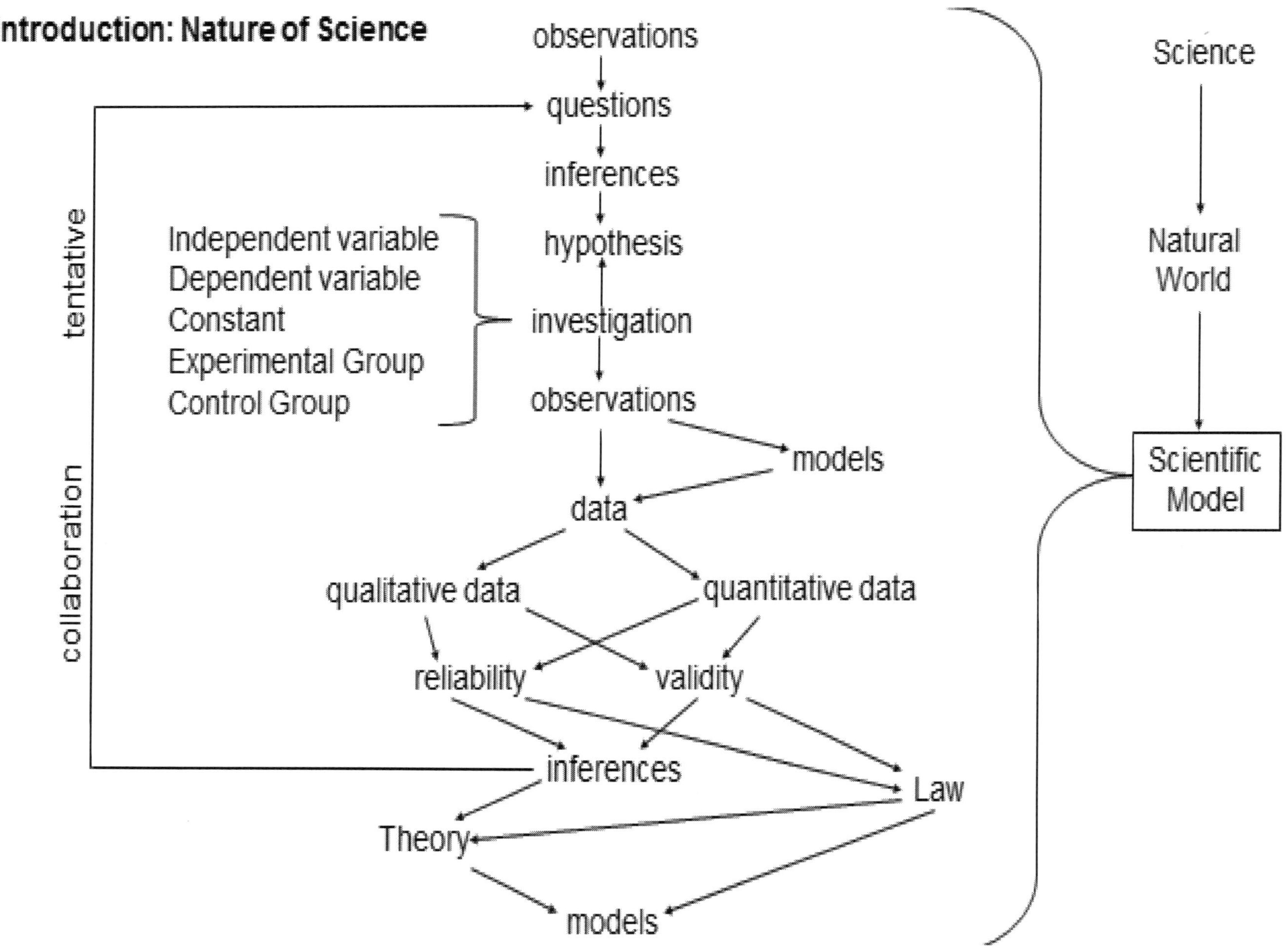
Introduction: Nature of Science
observations
questions
inferences
hypothesis
investigation
observations
models
data
qualitative data
quantitative data
reliability
validity
inferences
Law
Theory
models
Independent variable
Dependent variable
Constant
Experimental Group
Control Group
tentative
collaboration
Science
Natural World
Scientific Model

Focus on the Process

Directions:

Get a **small Legos set**. Teachers, make sure it is not too easy for your students. You are going to try to put it together in two different ways. Time how long it takes to put it together each way and answer the questions that follow. **Looking at the materials and lab we will be using, what are the safety precautions we should take to protect ourselves and materials during the investigation?**

A) Take the Lego pieces and construct the picture (the **product**) on the box's cover, looking at nothing but the cover and the Lego pieces.

B) When 20 minutes have passed, or you are done, take what you have made totally apart. Take out the directions (the **process**) and construct the product while using the step-by-step directions. Time how long it took you to complete the set.

Questions:

1) How did it feel trying to construct the Legos (A) without any directions?

2) Did you finish? If so, how long did it take?

3) How did it feel to construct the Legos (B) with the step-by-step directions?

4) Did you finish? If so, how long did it take?

5) Which strategy (A or B) allowed you to complete the product?

6) Which strategy (A or B) was more intimidating?

7) Which strategy (A or B) allowed you to see what is under the surface?

8) Which strategy (A or B) will allow you to learn more?

We often get anxious or procrastinate when faced with a large task. We are tempted to take a "shortcut" (copy or cheat, we do not learn much when we do this). There are pain and stress hormones that are released when this happens. One way to overcome this is to just worry about the next step in the process and not worry about the product. You can see and measure progress, which makes the process not feel too bad. Another way is just to start working. When you start working, those pain and stress hormones stop getting released so that anxiety goes away; this is why when we want to learn efficiently and effectively, we must:

Focus on the ____________________ and the ____________________ will take care of itself.

9) How is putting the Lego pieces together like putting ideas together to understand concepts?

Measurement Lab

Directions:

You will need **water**, a **scale**, a **meter stick**, a **temperature probe** attached to an **interface** connected to a **computer** with **Logger Pro**, a **100 mL graduated cylinder**, and a **stopwatch**. **Looking at the materials and lab we will be using, what are the safety precautions we should take to protect ourselves and materials during the investigation?**

1) Take the graduated cylinder and find its mass empty; write this in Data Table 1.
2) Add 50 mL of water to the graduated cylinder. Make sure you use the meniscus properly where the volume is at the bottom of the meniscus. Have the teacher check that you measured it correctly. Have each person in your group empty and fill the graduated cylinder with 50 mL of water. As they do so, have each person in your group time how long it takes for each person to fill the graduated cylinder and check it is correct (it is not a race, just a chance to get familiar with using the graduated cylinder and stopwatch).
3) Now find the mass of the graduated cylinder with 50 mL of water in it. Subtract the mass of the empty graduated cylinder from this mass and write the water's mass in Data Table 1.
4) Connect your temperature probe to an interface and connect your interface to a computer with Logger Pro (unless you have a LabQuest 2, then just hook your probe to the LabQuest 2). Find where the Logger Pro is located on your computer so you can use it again in the future. Once open, find the graduated cylinder's water temperature in Fahrenheit and Celsius (you will have to figure out how to change units). Write these in Data Table 1.
5) Take your meter stick and measure the length of the graduated cylinder. And measure the width of the base in centimeters. Write these in Data Table 1

Data Table 1

Object	Mass (g)	Volume (mL)	Time to Fill (s)	Temp (°F)	Temp (°C)	Length (cm)	Width (cm)
Graduated Cylinder		✖		✖	✖		
Water			✖			✖	✖

Questions:

1) Convert a length to meters, the volume to liters and a mass to kilograms, and Celsius to Kelvin.

   Length ________ m Volume ________ L Mass ________ kg Temp ________ K

2) What do you notice about the mass of the water compared to its volume?

3) What can happen to your investigations if your measurements are not accurate or precise?

4) Why do you think the rest of the world uses the metric system over the English system.

Patterns in Pennies

Directions:

You will need a **ruler**, 10 **pennies**, a **balance**, a **roll of pennies**, and an **empty penny roll**. **Looking at the materials and lab we will be using, what are the safety precautions we should take to protect ourselves and materials during the investigation?**

1) Find the mass of one penny with a scale to the nearest .1 g. Then measure the height of the penny in millimeters. Write these in Data Table 1 below.
2) Place another penny on top of the original penny and find the mass and height of the two pennies. Write these in Data Table 1 below.
3) Keep adding pennies one by one, measuring the mass and height until you have 10 pennies on the scale.
4) Make a line graph with the mass on the (*x*) axis and the height on the (*y*) axis for the pennies on Graph 1.

Data Table 1

Number of Pennies	Mass	Height
1		
2		
3		
4		
5		
6		
7		
8		
9		
10		

Graph 1

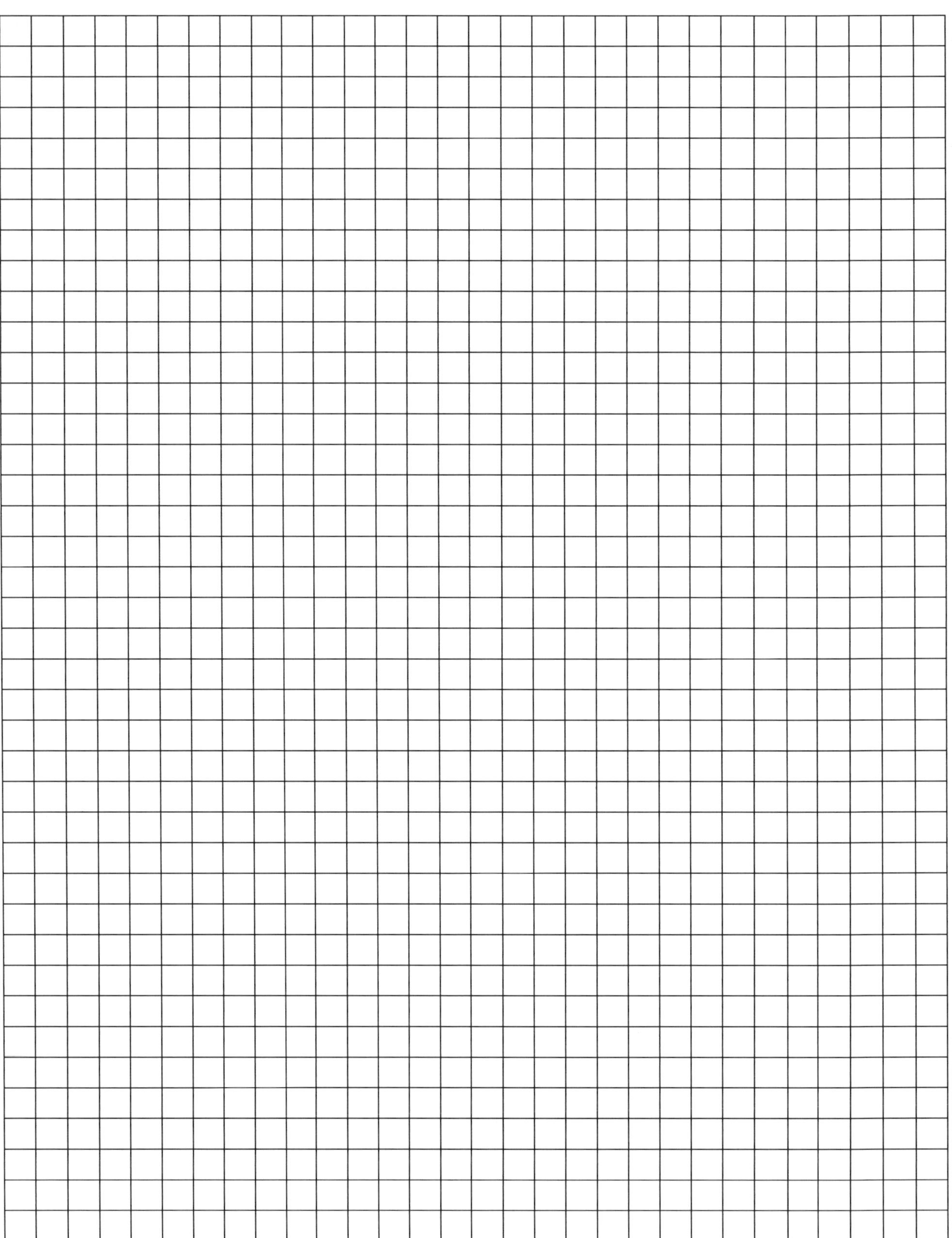

Questions:

1) What do you notice about the graph?

2) Is this a direct or inverse relationship between mass and height?

3) Do all pennies have the same mass? (Explain)

4) Do all the pennies have the same thickness? (Explain)

5) Use your data to estimate how many pennies are in the coin roll. How many pennies do you think are in the roll?

6) What did you do to estimate the number of coins?

7) What else could you do to estimate the coins?

8) Try your answer to #7. Do you get the same number as #5?

9) Carefully open up the coin roll and find out how many pennies there are. How close were you to the real number? After you are done counting, carefully close the roll back up.

10) Calculate the % accuracy by taking the lowest number between your guess and the actual number dividing by the higher of the two, then multiplying by 100.

11) What were some sources of error?

Virtual Investigations that go with Introduction

ExploreLearning.com

Unit Conversions Gizmo

Graphing Skills Gizmo

Measuring Volume Gizmo

Elevator Operator (Line Graphs) Gizmo

Weight and Mass Gizmo

Triple Beam Balance Gizmo

Reaction Time 1 Gizmo

Reaction Time 2 Gizmo

Physicsclassroom.com/Concept-Builders/Chemistry:

Measurement and Numbers

Significant Digits and Measurements

Metric System

Metric Estimation

Experiments and Variables

Proportional Reasoning

Calculating Slope

Using Graphs

Which One Doesn't Belong

Unit 1: Elements Compounds and Mixtures

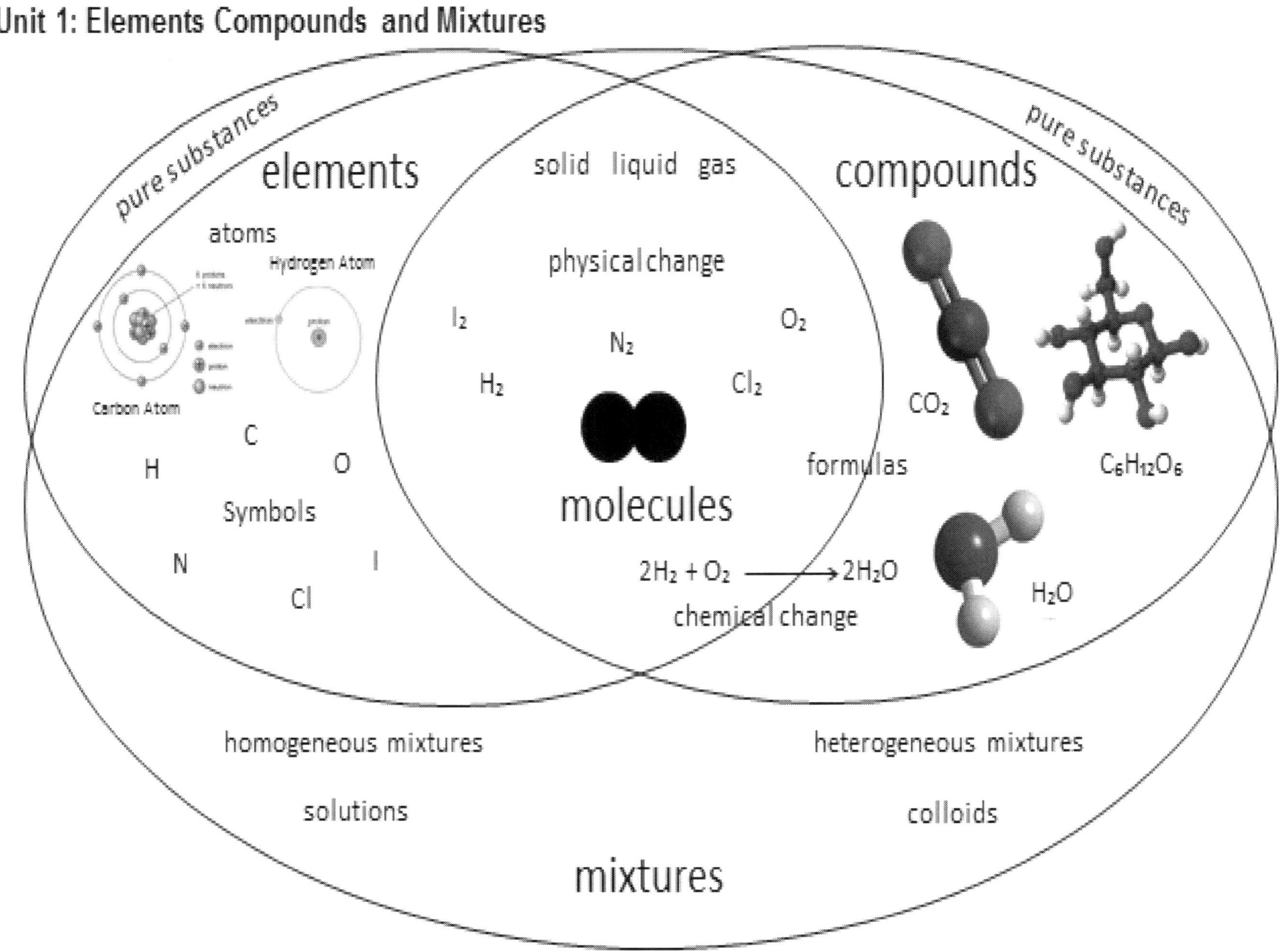

Metal or Nonmetal

Directions:

You will need samples of **sulfur**, **charcoal**, **copper**, **aluminum foil**, **mechanical pencil lead** (graphite), a **battery pack**, **batteries**, and a **Christmas light** with the ends of the wires stripped of insulation. **Looking at the materials and lab we will be using, what are the safety precautions we should take to protect ourselves and materials during the investigation?**

1) **Metals** are shiny, malleable, ductile, and conduct heat and electricity well. **Nonmetals** are brittle, dull, and do not conduct heat and electricity well.
2) Observe the five samples and fill in Data Table 1 below on those materials. Use the battery pack with batteries and the Christmas light to see if the materials conduct enough electricity to light the Christmas light when put into a circuit.

Data Table 1

Sample	Shiny or Dull	Brittle or Malleable	Conduct Electricity?	Metal or Nonmetal?
Sulfur				
Charcoal				
Copper				
Aluminum Foil				
Mechanical Pencil Lead				

Questions:

1) What are the characteristics of metals?
2) What are the characteristics of nonmetals?
3) Look at the periodic table and find carbon. Why do you think graphite conducted electricity?

Periodic Table Activity

Directions:

Using your **Periodic Table**, color the following with **colored pencils**:

1) Nonmetals are (yellow)
2) Metalloids are (green)
3) Metals are (blue)
4) Outline the alkali metals in (red)
5) Outline the alkaline earth metals in (black)
6) Outline the transition metals in (brown)
7) Outline the halogens in (blue)
8) Outline the noble gasses in (purple)

Add the following to the Periodic Table:

1) Atomic numbers
2) Group/family numbers
3) Periods/energy levels
4) Oxidation numbers
5) Label the Lanthanide series and Actinide series

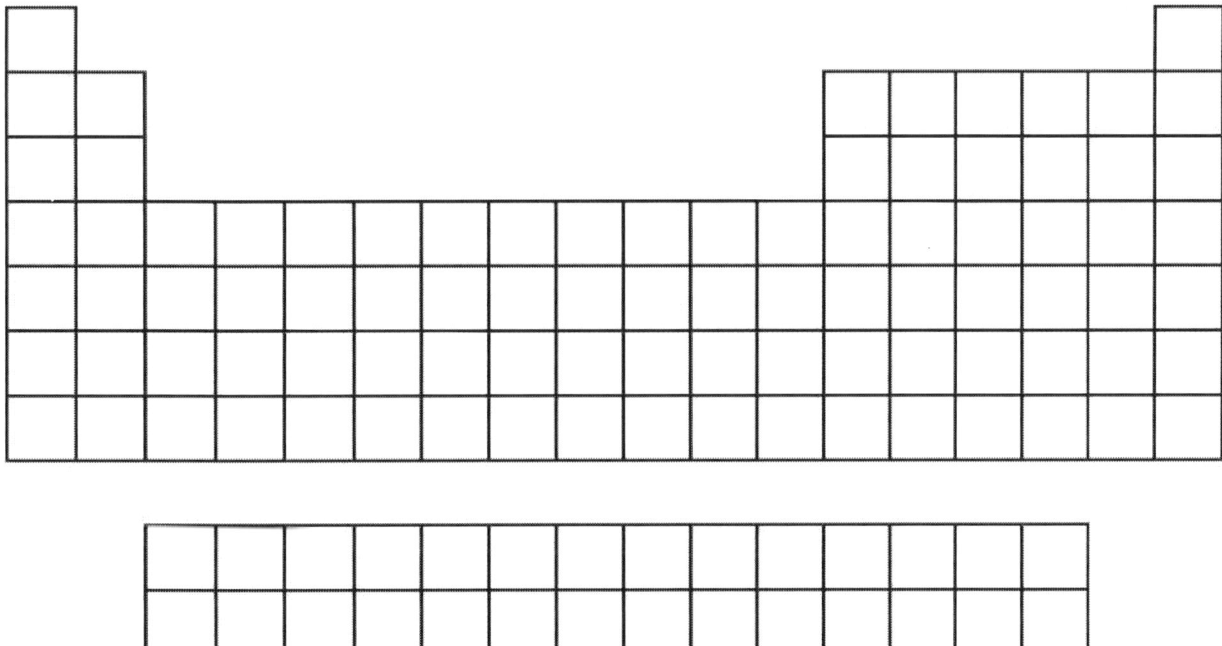

Questions:

1) What do groups have in common?

2) What do periods have in common?

3) Where are the most reactive nonmetals?

4) Which group/family does not react at all because they are full?

5) If you can't remember from previous investigations, research the internet to find these three things:
    a. What are the characteristics of metals?
    b. What are the characteristics of nonmetals?
    c. What are the characteristics of metalloids?

Elements Compounds and Mixtures Research

Directions:

Using your teacher's instructions, use the **internet** and your **textbook** to research elements, compounds, and mixtures, then answer the following questions.

1) What is an element, and how is it related to compounds and mixtures?

2) What are examples of elements?

3) What is a compound, and how is it related to elements and mixtures?

4) What are examples of compounds?

5) What is a mixture, and how is it related to elements and compounds?

6) What are the different categories of mixtures, and how do you tell them apart?

7) How were the different elements made?

8) Can an element be separated?

    a. Is it easy or hard? Explain.

9) How can a compound be separated?

    a. Is it easy or hard? Explain.

10) How can a mixture be separated?

    a. Is it easy or hard? Explain.

11) What does this investigation tell us about ourselves?

12) Fill in the Ven diagram below comparing and contrasting elements, compounds, and mixtures.

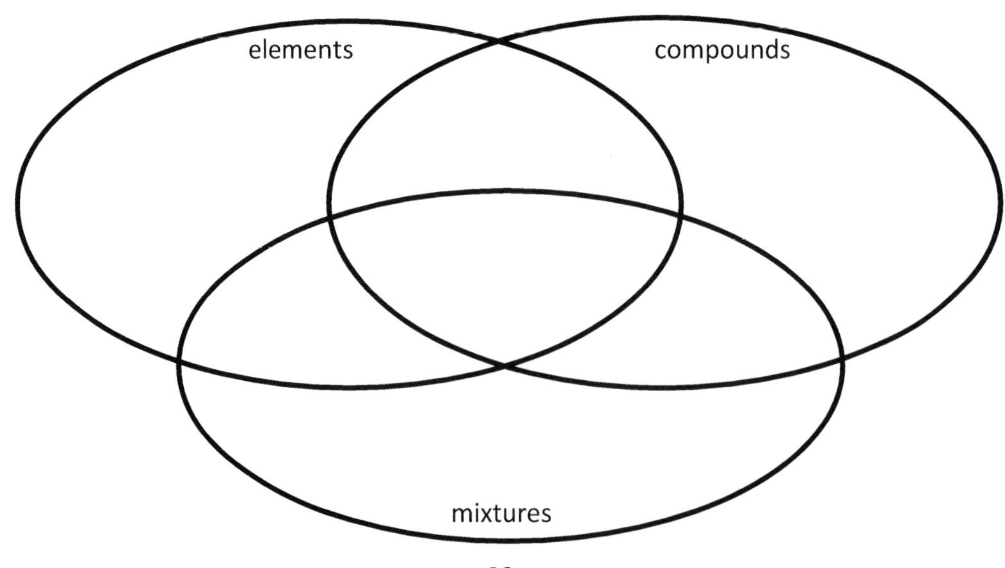

Elements Compounds and Mixtures

Directions:

You will need **aluminum foil, milk**, a **laser pointer, water, granite countertop samples, Kool-Aid, salt** (sodium chloride), a **pencil, chalk** (calcium carbonate), and **muddy water. Looking at the materials and lab we will be using, what are the safety precautions we should take to protect ourselves and materials during the investigation?**

1) **Pure substances** can be **elements** or **compounds**.
    a. **Elements** are pure substances with only one type of atom found on the periodic table.
    b. **Compounds** are pure substances with two or more elements in a fixed ratio.
2) **Mixtures** have varying ratios of different substances.
    a. **Homogeneous mixtures** appear the same throughout. They are also called **solutions**. Lasers can go through **solutions**.
    b. **Heterogeneous mixtures** are different throughout. Lasers are deflected in heterogeneous mixtures.
        i. **Colloids** look like homogeneous mixtures but have large particles that do not settle out.
        ii. **Suspensions** have larger particles that do settle out.
3) Analyze each substance in Data Table 1 and classify it as either a pure substance or a mixture. Then classify the type of pure substance or mixture. This data will be filled in Data Table 1.

Data Table 1

Object	Identity (Pure or Mix)	Classification
Aluminum foil		
Milk		
Water		
Salt		
Granite countertop		
Cool aid		
Salt		
Pencil		
Chalk		
Muddy water		

Questions:

1) If you know the name of a substance, how can you tell whether it is an element or not?

2) How did you find out how milk was classified?

3) How would you classify a human? Explain why.

4) How would you classify fog?

5) How would you classify a clean atmosphere?

6) How would you classify smog?

7) How would you classify brass?

Separating Mixtures

Directions and Questions:

You will need a **magnet** in a **plastic baggy** (keep the magnet in the plastic baggy the whole time), a **wire strainer**, a **coffee filter**, a **hotplate**, **water**, **sand**, **sugar**, **marbles**, **iron filings**, **granola**, and two **beakers**. **Looking at the materials and lab we will be using, what are the safety precautions we should take to protect ourselves and materials during the investigation?**

1) Take the mixture of sand, sugar, marbles, iron filings, and granola and use the materials to separate the mixture into its pieces. How will you divide the marbles from the mixture?

2) How will you separate the iron filings from the mixture?

3) How will you separate the granola from the mixture?

4) How will you separate the sand from the mixture?

5) How will you separate the sugar from the mixture?

6) Was this a homogeneous or heterogeneous mixture?

Separating Pigments

Directions:

You will need **goggles**, **scissors**, different **pens** and **markers**, an **eyedropper**, **nail polish remover** or **alcohol**, **filter paper** or **chromatography paper**, **test tubes**, a **test tube rack**, **paper clips**, and **rubber stoppers** with holes in them that fit the test tubes. **Looking at the materials and lab we will be using, what are the safety precautions we should take to protect ourselves and materials during the investigation?**

1) You are going to do chromatography today. Take the paper and cut it into strips if you have to, with a point at one end.
2) Make a dot with the pen or marker about an inch from the pointy tip.
3) Take a paper clip and bend it so there is a hook on one end that you poke through the flat end of the strip of paper and put the other end through the hole in the stopper.
4) Take the eyedropper and carefully squirt a little nail polish remover/alcohol into the bottom of a test tube. Then lower the paper with the dot pointy end down into the nail polish remover, but do not let the dot touch it. Fix the stopper at the top of the test tube. Bend the part of the paper clip that is above the stopper, so the paper does not drop any lower.
5) Repeat the process you did for #s 1-4 for all the different pens and markers you have chosen.
6) Watch the patterns of pigments that separate. The ones that move the fastest go to the top, and those that move the slowest will be at the bottom.
7) After your pigments have separated, take them out of the test tubes and let them dry out (otherwise, all the stains may go back together again at the top of the paper).
8) Draw pictures of the color bands that came out from your pens and markers on your paper strips below:

Questions:

1) Did you see the same colors make different patterns?

2) How could you use this to find out which pen wrote a note if you needed to?

3) Classify the matter of the ink as an element, compound, homogeneous mixture, or heterogeneous mixture. Explain why.

4) Which careers might be interested in this technique for separating pigments?

50 + 50 Does Not Equal 100

Directions and Questions:

You will need **safety goggles**, two **400 mL beakers**, two **100 mL graduated cylinders, sand, marbles**, 50 mL of **water**, and 50 mL of **rubbing alcohol. Looking at the materials and lab we will be using, what are the safety precautions we should take to protect ourselves and materials during the investigation?**

1) Carefully measure 50 mL of water in one graduated cylinder and 50 mL of alcohol in another.
2) Carefully pour the contents of one of the graduated cylinders into the other. What is the total volume now?
3) How do you think this could have happened?
4) Fill a beaker 200 mL full with marbles and another 200 mL full with sand. Gently pour the sand into the marbles, slowly shaking the beaker of the marbles. How many milliliters of marbles and sand is there now in the beaker?
5) Where did the sand go to lose the volume?
6) How can you explain the results for #2 now?

Percent Sugar in Bubble Gum

Directions and Questions:

You will need a **scale** and **bubble gum**. **Looking at the materials and lab we will be using, what are the safety precautions we should take to protect ourselves and materials during the investigation?**

1) Find the mass of one piece of gum sitting on the wrapper; write this in Data Table 1.
2) Take the gum and chew it. As you are chewing, find the mass of the wrapper and write it in Data Table 1.
3) Subtract the wrapper's mass from the gum and wrapper's mass to get just the unchewed gum's mass; write this in Data Table 1.
4) When you notice no more flavor in the gum, put the chewed gum back on the wrapper and find the chewed gum's mass (do not forget to subtract the wrapper's mass). Write the mass of the chewed gum in Data Table 1.
5) To find the percent of the gum that was not sugar, take the chewed gum's mass and divide it by the unchewed gum's mass times 100.

Data Table 1

Mass of unchewed gum and wrapper	Mass of wrapper	Mass of unchewed gum	Mass of chewed Gum	% of gum, not sugar	% of gum that is sugar
g	g	g	g	g	g

Questions:

1) What was the percentage of gum that is sugar?

2) Where did the sugar and flavor go?

3) So when you are chewing gum, are you technically eating? Explain why.

4) Classify gum as an element, compound, homogeneous mixture, or heterogeneous mixture. Explain why.

5) Which careers might be interested in this type of investigation?

Extension: try this for different brands of gum or even sugarless gum. Show your results below.

Virtual Investigations that go with Elements Compounds and Mixtures

ExploreLearning.com:

Element Builder

Average Atomic Mass

Chemical Equations

Isotopes

Ionic Bonds

Periodic Trends

Polarity and Intermolecular Forces

Collision Theory

Covalent Bonds

Melting Points

Chemical and Physical Changes STEM Case

Chemical and Physical changes Handbook

Properties of Matter STEM Case

Properties of Matter Handbook

Phet.colorado.edu:

Build an Atom

Build a Molecule

Isotopes and Atomic Mass

Atomic Interaction

Beer's Law Lab

Concentration

Ration and Proportion

Molarity

Molecular Polarity

Molecule Shapes

Molecule Shapes: Basics

Physicsclassroom.com/Concept-Builders/Chemistry:

Classification of Matter

Name that Element

Particles..Words..Formulas

Subatomic Particles

Isotopes

Atomic Models

Ionic Bonds

Bond Polarity

Molecular Polarity

Unit 2: Properties of Water

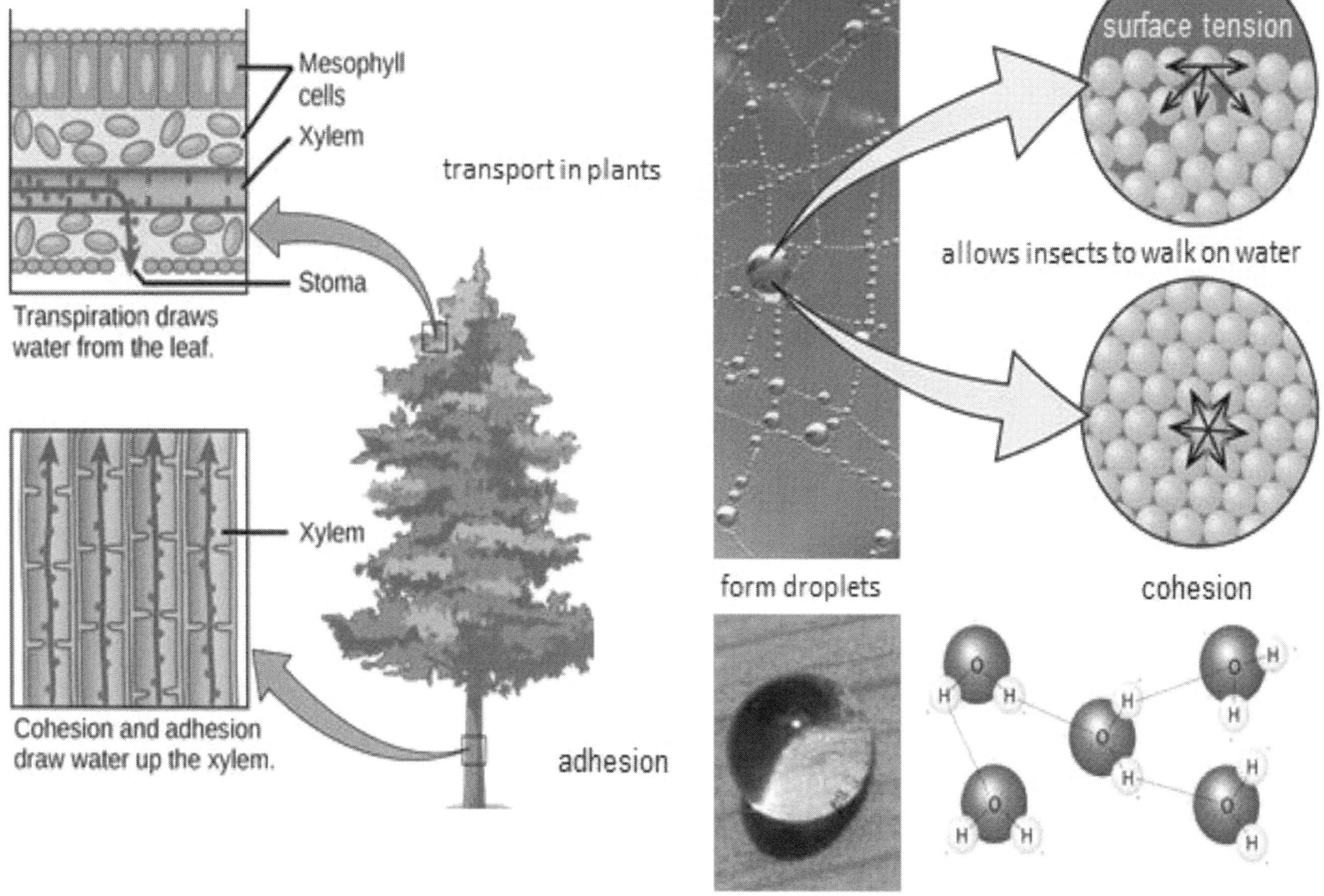

Mesophyll cells

Xylem

Stoma

Transpiration draws water from the leaf.

Xylem

Cohesion and adhesion draw water up the xylem.

transport in plants

adhesion

surface tension

allows insects to walk on water

form droplets

cohesion

Building a Model of a Water Molecule

Directions:

You will need a **balloon**, a **molecular model kit,** and a **Periodic Table**. **Looking at the materials and lab we will be using, what are the safety precautions we should take to protect ourselves and materials during the investigation?**

1) At the top of your periodic table, label it like this just below:

+1 +2 skip the transition metals then +3 +4- -3 -2 -1 0

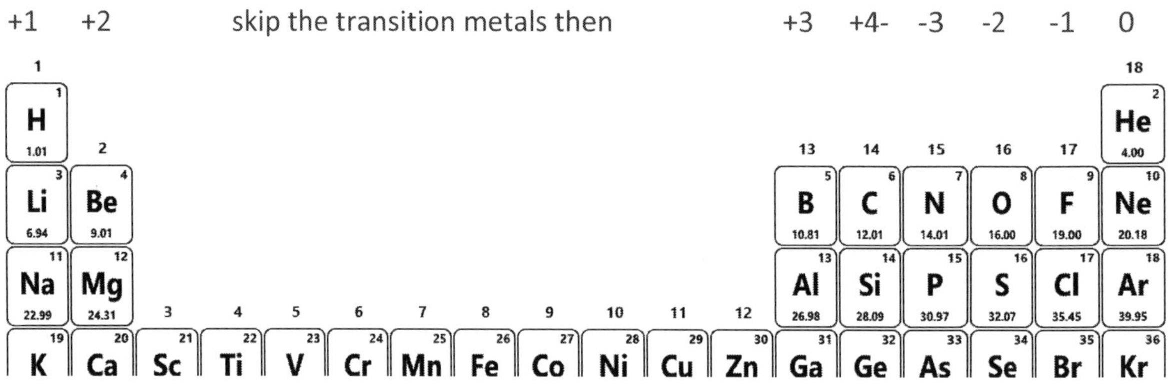

2) Different kits have different colors. In my kit, the:
    a. +1 (one-prong white) represents the Alkali Metals
    b. +2 (two-prong yellow) represents the Alkaline Earth Metals
    c. +3 (three-prong blue) represents the Boron Group
    d. +/- 4 (four-prong black) represents the Carbon Group
    e. -3 (three-prong red) represents the Nitrogen Group
    f. -2 (two-prong blue) represents the Oxygen Group
    g. -1 (one-prong green) represents the Halogens

3) Use the pieces to make two H_2O molecules. The hydrogen side of the molecule is slightly positive, and the oxygen side of the molecule is slightly negative making it polar like a magnet.

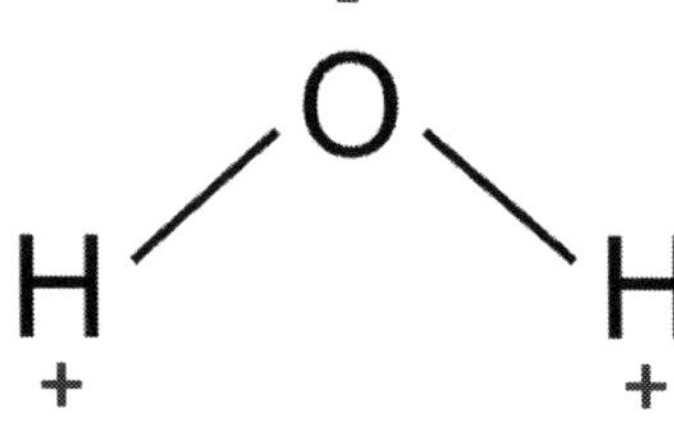

4) Because the water molecule has positive and negative ends, **ions** are attracted to the opposite charges on the water molecule. The positive ions are attracted to the oxygen side, and the negative ions are attracted to the hydrogen side. The same is true for other **polar molecules**; this is why ionic compounds and polar molecules like to dissolve in water. We call water the **universal solvent**.

5) Make a model of liquid water by taking your two water molecules and placing them next to each other where the oxygen of one is sitting between the two hydrogens of the other; this is how water molecules like to stick to each other. The positive ends are attracted to the negative ends; this is why water is **cohesive** (it sticks together).

6) When there are a bunch of them together, they have an equal pull on each other except for the ones on the surface; they are pulled slightly down because they have a slight charge above them. After all, there are no other molecules above them; this is why water has **surface tension**.

7) Since it is charged on both ends, it is also attracted to surfaces like a balloon with a static charge is attracted to a sweater or a wall. This attraction is why see water clinging to the sides of cold cans or glasses of ice tea. This phenomenon is called **adhesion**.
    a. You can model this by taking an inflated balloon, rubbing it on your hair to steal some electrons, and then sticking it to a shirt or wall.

8) You can make a model of ice (solid water) by flipping one of your water molecules and facing the oxygen ends toward each other. When water gets cold, the molecule's charge is not as strong, and the opposite ends are not attracted to each other anymore, so the oxygen atoms come together and share electrons with each other bonding them together. This orientation gives the molecule more space inside it and is why ice floats in liquid water.

Question:

1) Why do you think life depends so much on water?

2) How did these models show the characteristics of water?

3) How were the models inaccurate?

Checking Polarity

Directions:

You will need two **pennies** tails up on a **paper towel**, two **pipettes** or **eyedroppers**, **water**, and **rubbing alcohol**. **Looking at the materials and lab we will be using, what are the safety precautions we should take to protect ourselves and materials during the investigation?**

1) With a pipet or eyedropper, slowly place drops of water on one penny. Count how many drops you could put on the penny before it spilled off. Write that number in Data Table 1 below.
2) With a different pipet/eyedropper, slowly place drops of rubbing alcohol on the second penny. Count how many drops you could put on the penny before it spilled off. Write that number in Data Table 1.

Data Table 1

Substance Dropped	Number of Drops
Water	
Alcohol	

Questions:

1) Which substance was able to have the most drops put onto the penny?

2) Polarity is in a substance with opposite charges at its ends. Which substance had more polarity?

3) How is a polar molecule like a magnet?

4) **Cohesion** is when water molecules stick to other water molecules. How does the polarity work to allow the water drop to hold together?

    a. Where did we see this in the investigation?

5) **Adhesion** is where water molecules stick to other substances. Where did we see this in the investigation?

6) **Surface tension** is the pulling of the surface water molecules down to the rest of the water because no other force is pulling them up or away. Where did we see surface tension in the investigation?

    a. How do you think this would allow insects to walk on water?

7) How do you think water can defy gravity by moving up the stems of plants?

Celery Transport

Directions and Questions:

You will need fresh **celery** with leaves, a **beaker**, and **red** and **blue food coloring**. This investigation will also work with a **white carnation** and **all colors of food coloring. Looking at the materials and lab we will be using, what are the safety precautions we should take to protect ourselves and materials during the investigation?**

1) In a beaker of water, add a couple of drops of food coloring. Put the celery into the water with the leaves up out of the water. What do you think will happen with the water in the beaker and the celery? Hypothesis:

2) Let the water and celery sit overnight. Come back the next couple of days and observe what you see. What did you see in the celery's leaves?

3) What do you think caused the celery to look like this?

4) How is the xylem in the celery stem like a straw in a drink?

5) Discuss with your teacher and the class how pressure and magnetism were involved with the process you observed in this investigation. Explain what you discussed.

6) Is pressure or magnetism involved with **cohesion** and **adhesion**? Explain.

7) **Transpiration** is the process of water moving up the plant from the roots to the leaves through xylem tubes. How is **cohesion** involved with transpiration?

8) How is **adhesion** involved with transpiration?

Transpiration Pull

Directions and Questions:

You will need a **pressure sensor** and the **tube setup** attached to an **interface** connected to a **computer** with **Logger Pro**. You will then need to find a **plant branch** that will snuggly fit inside the tube, sealing the tube. **Looking at the material and lab we will be using, what are the safety precautions we should take to protect ourselves and materials during the investigation?**

1) Once your pressure sensor is connected to the interface and the computer with the Logger Pro, set the data collection to collect data for 5 minutes. Put the plant branch inside the tube connected to the pressure sensor to seal it.
2) Press "Collect" on the Logger Pro. Watch the data for a few minutes. What do you see happening to the measurement of the pressure sensor?

3) Why do you think this is happening?

4) When the data seems to level out, break the seal, let the pressure equalize with the atmosphere, and stop the data collection if it has not already stopped. Set up the experiment to run again. Did you see the same trend in the results?

5) How does this show evidence of transpiration pull?

Seeing a Stoma

Directions and Questions:

You will need a **textbook**, **clear scotch tape**, **lettuce**, a **slide**, a **compound light microscope**, and the **internet** or **textbook**. **Looking at the materials and lab we will be using, what are the safety precautions we should take to protect ourselves and materials during the investigation?**

1) Take a small piece of scotch tape, put the sticky side on the lettuce, and then peel it off. You should have just removed one layer of cells from the outside of the lettuce. You should see the epidermis, which contains the guard cells and stomata.
2) Place the sticky side of the tape down on the slide. Place the slide on the microscope stage and follow your teacher's instructions on centering and focusing it. Draw a picture of the lettuce epidermis labeling the guard cells and stoma.

3) Research how the guard cells open and close the stoma and describe it here:

4) Why does the rest of the epidermis look like puzzle pieces?

5) What is the function of the epidermis?

6) What is the function of a stoma?

7) How does a stoma allow water to move up the stem?

8) How do cohesion and adhesion help move water up the stem of a plant?

9) What would happen if all the stomata would close in a plant, would transport up the stem be possible? Explain why.

How does Rain Form?

Directions and Questions:

You will need a **glass** or **beaker of ice water**. **Looking at the materials and lab we will be using, what are the safety precautions we should take to protect ourselves and materials during the investigation?**

1) Why do you see water forming on the outside of the glass?

2) How is this like water forming droplets in the sky, making clouds and rain?

    a. What is the difference between clouds and rain?

3) How is a liquid different from a gas allowing this to happen with water?

4) What allows the water to stick together on the glass?

5) What holds the water drops to the glass?

6) How are **cohesion**, **adhesion**, and **surface tension** involved?

7) If an animal was small enough, explain how it could walk on water.

Virtual Investigations that go with Properties of Water

ExploreLearning.com:

Water Cycle

Freezing Point of Salt Water

Phases of Water

Phase Changes

Sticky Molecules

Solubility and Temperature

Colligative Properties

Covalent Bonds

Polarity and Intermolecular Forces

Phet.colorado.edu:

Molecule Polarity

Molecule Shapes

Molecule Shapes: Basics

States of Matter

States of Matter: Basics

Balloons and Static Electricity

Physicsclassroom.com/Concept-Builders/Chemistry:

Bond Polarity

Molecular Polarity

Unit 3: Acids & Bases

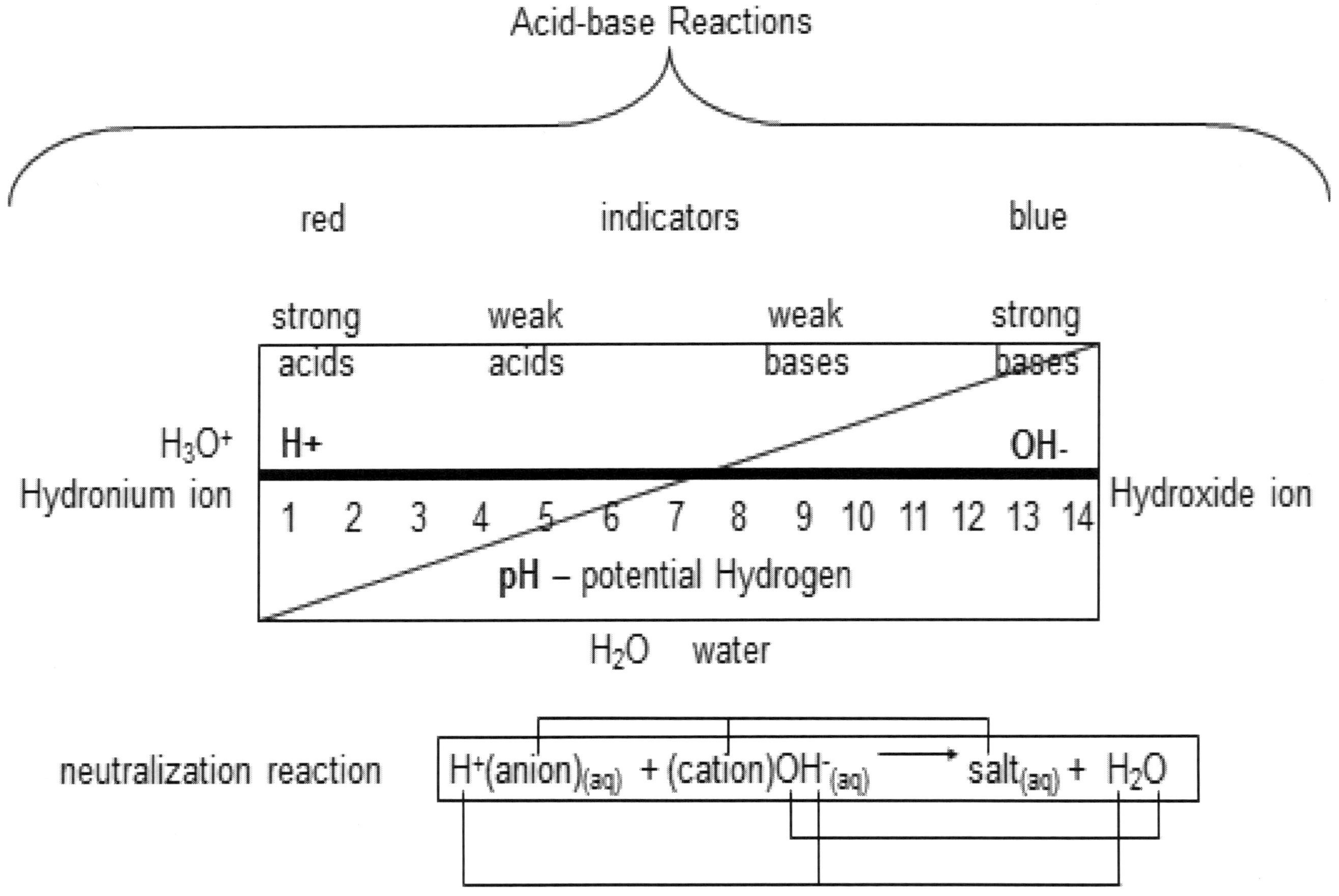

Acid-base Reactions

red | indicators | blue

strong acids	weak acids	weak bases	strong bases

H_3O^+ Hydronium ion — H+ … OH- — Hydroxide ion

1 2 3 4 5 6 7 8 9 10 11 12 13 14

pH – potential Hydrogen

H_2O water

neutralization reaction

$H^+(anion)_{(aq)} + (cation)OH^-_{(aq)} \longrightarrow salt_{(aq)} + H_2O$

Which is an Acid and Which is a Base?

Directions:

Acids have H+ ions, and bases have OH- ions. Look at the formulas of some common acids and bases and see if you can tell them apart. Notice positive ions are written on the left side of the formula, and negative ions are written on the right side of the formula. Next to each formula, write down whether you think it is an acid or a base.

1) $Ca(OH)_2$
2) H_2SO_4
3) HCl
4) NaOH
5) $HC_2H_3O_2$
6) NH_4OH
7) KOH
8) H_2CO_3
9) Ba $(OH)_2$
10) CsOH
11) HNO_3
12) HBr
13) $Mg(OH)_2$
14) HI
15) LiOH
16) $Al(OH)_3$
17) RbOH
18) $HClO_4$
19) $HClO_3$
20) $Fe(OH)_2$
21) HBO_3
22) H_2S
23) $Zn(OH)_2$
24) HNO_2
25) $Fe(OH)_3$
26) HPO_3
27) $HC_2H_3O_2$
28) H_2SiO_3
29) How did you distinguish between an acid and a base with these compounds?

A Homemade Indicator

Directions:

You will need **safety goggles**, an **apron**, a **500 mL beaker**, five **small beakers**, **water**, **red cabbage**, a **hotplate**, **shampoo**, **grapefruit juice**, **Sprite**, **milk**, and **ammonia**. **Looking at the materials and lab we will be using, what are the safety precautions we should take to protect ourselves and materials during the investigation?**

1) Place about 300 mL of water in the 500 mL beaker with some leaves of purple cabbage. Boil it until the water turns purple.
2) In 5 small beakers, separately place shampoo, grapefruit juice, Sprite, milk, and ammonia.
3) The red cabbage juice is red in most acids and blue-purple in most bases. Use a pipet to mix the red cabbage juice with each substance to determine whether it is **acid** or a **base**.

Results:

4) What is shampoo?

5) What is grapefruit juice?

6) What is Sprite?

7) What is milk?

8) What is ammonia?

Observing Acid Relief

Directions and Questions:

You will need **100% purple grape juice**, a **beaker**, **water**, and an **antacid tablet**. **Looking at the materials and lab we will be using, what are the safety precautions we should take to protect ourselves and materials during this investigation?**

1) Take a beaker and fill it halfway up with a solution of water and purple grape juice. The pigment in grape juice is a natural indicator like red cabbage juice. It turns red in acids and a grayish blue in a base. What color is the grape juice in the water solution?

2) Is the solution an acid or base?

3) When stomach acid gets too strong, we take antacid tablets to calm our stomach down. Drop an antacid tablet into the water and grape juice solution. What do you notice happening?

4) Is the ending solution an acid or base?

5) How does an antacid tablet help calm our stomach or acid reflux?

6) Which careers might be interested in this type of reaction?

Acid or Base Grape Juice Indicator

Directions:

You will need **safety goggles**, an **apron**, a **pipette**, **100% purple grape juice**, seven tiny **beakers**, **vinegar**, **ammonia**, **lemon juice**, **Sprite**, **drain cleaner**, **detergent**, a **baking soda solution**, **litmus blue paper**, **litmus red paper**, **Universal indicator pH paper**, and a **pH meter** attached to an **interface** connected to a **computer** with **Logger Pro**. **Looking at the materials and lab we will be using, what are the safety precautions we should take to protect ourselves and materials during the investigation?**

1) Fill each of the seven beakers with one of these substances: vinegar, ammonia, lemon juice, Sprite, drain cleaner, detergent, and baking soda.
2) Use the red and blue litmus papers, dip them into each beaker, and write down what color they turn in Data Table 1.
3) Use the universal indicator pH paper, dip them into each beaker, and write down what pH the color indicated in Data Table 1.
4) Gently stir the pH meter in each beaker to see what the Logger Pro indicates is the pH of each solution. Be patient; it takes some time for the pH meter to stabilize. Make sure to rinse the pH meter between each beaker measurement. Write the pH measurement in Data Table 1.
5) Now take a pipette full of purple grape juice, squirt it into each beaker, and write down the color it makes in each of the solutions in Data Table 1.

Data Table 1

Solution	Blue Litmus	Red Litmus	Universal Indicator Paper	pH meter	Purple Grape Juice	Acid or Base?
Vinegar						
Ammonia						
Lemon juice						
Sprite						
Drain cleaner						
Detergent						
Baking soda						

Questions:

1) A pH below 7 indicates an acid; a pH above 7 indicates a base. What color does litmus paper turn when a solution is an acid?

2) What color does litmus paper turn when a solution is a base?

3) Determine if each substance is an **acid** or a **base** by filling in the last column in Data Table 1.

4) Which of the substances was the strongest acid (had the lowest pH)?

5) Which of the substances was the strongest base (had the highest pH)?

6) What color did the purple grape juice turn in the solution if it was an acid?

7) What color did the purple grape juice turn the solutions if it was a base?

8) Which of the indicators we used was most like the grape juice indicator? Explain.

9) Which of the indicators told you the most information? Explain.

10) Why do you think the purple grape juice Is red when it is diluted?

Characteristics of Acids and Bases

Directions:

You will need **bottled water**, a **sink**, and put **vinegar**, **club soda**, **lemon juice**, **baking soda**, **soft soap**, and **laundry detergent** into **tiny cups** or **beakers**. **Looking at the materials and lab we will be using, what are the safety precautions we should take to protect ourselves and materials during the investigation?**

1) **Acids** are **sour,** and **bases** are **bitter** to the taste. **Bases feel slippery,** and **acids do not**. For each cup, separately dip your finger in and rub it on your thumb to find the feel (slippery or not slippery), and then lightly lick your finger to get a taste (bitter or sour). You can use the bottled water to wash your mouth out to get rid of the taste and use the sink to wash your fingers.
2) Circle the results in Data Table 1 below.

Data Table 1

Substance	Feel	Taste
Vinegar	Slippery or Not Slippery	Bitter or Sour
Club Soda	Slippery or Not Slippery	Bitter or Sour
Lemon Juice	Slippery or Not Slippery	Bitter or Sour
Baking Soda	Slippery or Not Slippery	Bitter or Sour
Soft Soap	Slippery or Not Slippery	Bitter or Sour
Laundry Detergent	Slippery or Not Slippery	Bitter or Sour

Questions:

1) Which substances were acids?

2) Which substances were bases?

Which will Corrode a Nail?

Directions:

You will need **two nails**, a small **bottle of coke** (you could also try **orange juice**), and a small bottle of **clear liquid soap**. **Looking at the materials and lab we will be using, what are the safety precautions we should take to protect ourselves and materials during the investigation?**

1) Acids tend to corrode metal, and bases do not. Coke has multiple acids in it, and soap is a base. **Hypothesis:** Which do you think will corrode the nail?

2) Open the lids of both the coke and the soap, place a nail in each one, and tightly fix the lids back on each.
3) Check them each day for two weeks to look for any signs of corrosion.
4) Fill in Data Table 1 below.

Data Table 1

Days	Corrosion of Nail in Coke	Corrosion of Nail in Soap
1	Yes or No	Yes or No
2	Yes or No	Yes or No
3	Yes or No	Yes or No
4	Yes or No	Yes or No
5	Yes or No	Yes or No
6	Yes or No	Yes or No
7	Yes or No	Yes or No
8	Yes or No	Yes or No
9	Yes or No	Yes or No
10	Yes or No	Yes or No
11	Yes or No	Yes or No
12	Yes or No	Yes or No
13	Yes or No	Yes or No
14	Yes or No	Yes or No

Question:

1) Did either show signs of corrosion in two weeks? If so, how?

2) Why do you think the pH of tap water is kept just over 7?

3) Why do you think soaps are good for cleaning metals?

4) Lemons contain lots of citric acid. Do you think lemon juice would corrode a nail? Explain.

Virtual Investigations that go with Acids and Bases

ExploreLearning.com:

pH Analysis Gizmo

pH Analysis: Quad Color Indicator Gizmo

Titration Gizmo

Phet.colorado.edu:

Acid-Base Solutions

pH Scale

pH Scale Basics

physicsclassroom.com/Concept-Builders/Chemistry:

Which One Doesn't Belong? Acid-Base Properties

Bronsted-Lowry Model of Acids and Bases

pH and pOH

Dissociation

Unit 4: Conservation of Mass

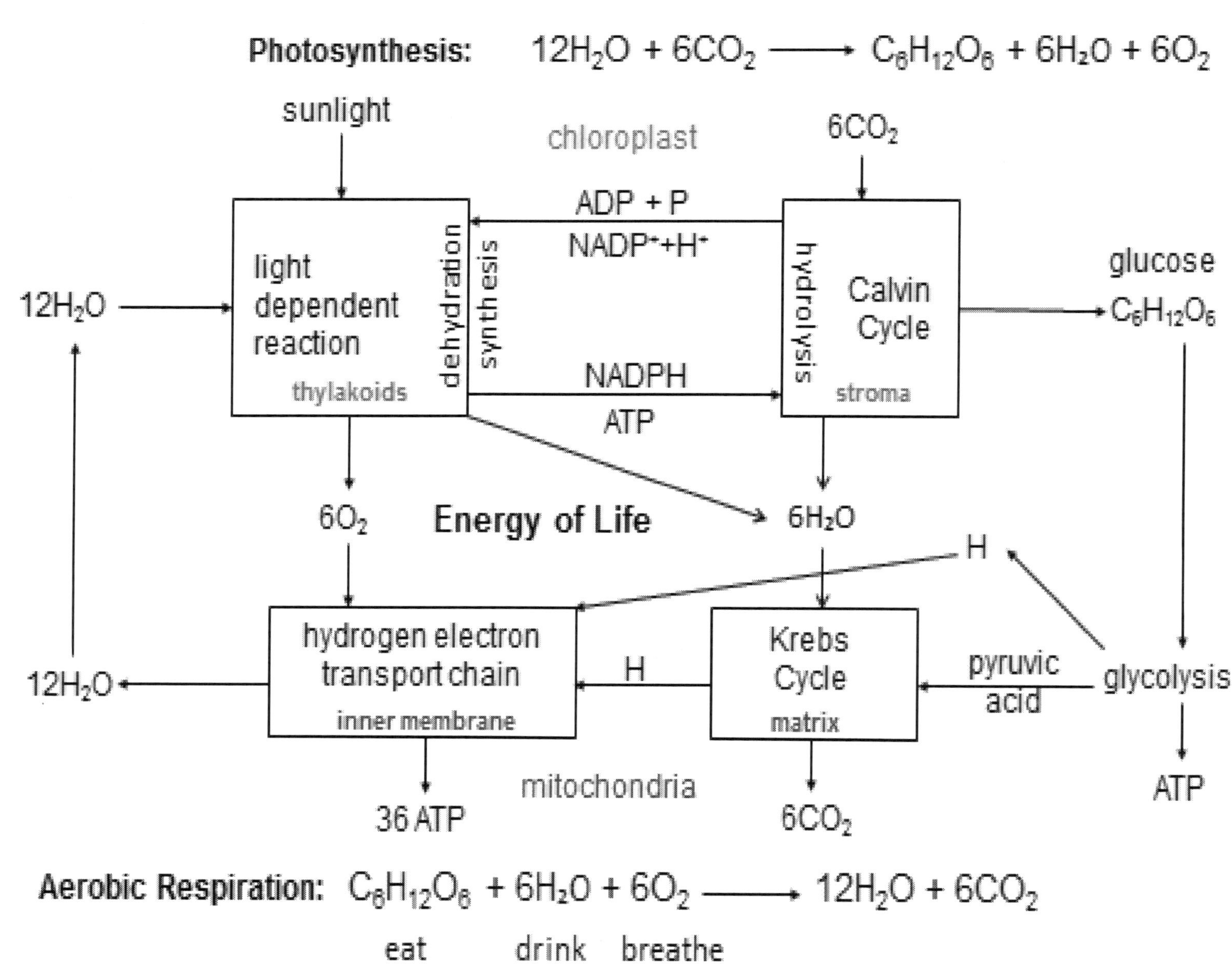

Conservation of Mass in Equations

Directions:

Draw lines from the elements in the reactants (the left side of the arrow) to the elements in the products (the right side of the arrow) for each balanced chemical equation to show how elements do not go away in a chemical reaction; they just get rearranged. An example is done below.

Example: $2H_2 + CO \rightarrow CH_3OH$

1) $2H_2O \rightarrow 2H_2 + O_2$

2) $HCl + NaHCO_3 \rightarrow CO_2 + H_2O + NaCl$

3) $C_6H_{12}O_6 + 6O_2 \rightarrow 6CO_2 + 6H_2O$

4) $CH_4 + 2O_2 \rightarrow CO_2 + 2H_2O$

5) $2H_2 + 2O_2 \rightarrow 2H_2O$

6) $C_2H_5OH + 3O_2 \rightarrow 2CO_2 + 3H_2O$

7) $6CO_2 + 6H_2O \rightarrow C_6H_{12}O_6 + 6O_2$

8) $2LiOH + CO_2 \rightarrow Li_2CO_3 + H_2O$

9) $(NH_4)_2Cr_2O_7 \rightarrow Cr_2O_3 + N_2 + 4H_2O$

10) $4NH_3 + 5O_2 \rightarrow 4NO + 6H_2O$

Questions:

1) How do these equations show mass is conserved?

2) How can a reaction have different reactants from its products but still conserve mass?

3) What do the numbers in front of the formulas represent?

4) How can you tell how many elements there are in an equation?

5) Which careers use the information in the conservation of mass?

Home Chemistry

Directions and Questions:

You will need **safety goggles**, **liver**, **hydrogen peroxide**, three small **cups**, **water**, **vinegar**, and **baking soda**. **Looking at the materials and lab we will be using, what are the safety precautions we should take to protect ourselves and materials during the investigation?**

1) All living things produce an enzyme called catalase. Catalase helps break down hydrogen peroxide into water and oxygen. Place a piece of liver in a cup and pour some hydrogen peroxide we use as an antiseptic on it. What do you observe?

    a. Write a balanced equation for the reaction.

2) One of the most useful natural gasses we use as fuel is methane (CH). It combines with oxygen in a combustion reaction to create heat, carbon dioxide, and water. Write a balanced equation for the reaction.

3) Place baking soda ($NaHCO_3$) and vinegar (CH_3COOH) together in a cup and watch the reaction. The products are carbon dioxide (CO_2), water (H_2O), and sodium acetate (CH_3COONa). What do you observe?

    a. Write a balanced equation for this reaction.

4) What do you think we use baking soda for when we cook?

Types of Chemical Reactions

Equipment and Safety:

You will need **safety goggles**, an **apron**, a **small aluminum pan**, a **digital scale**, **steel wool**, **baking soda**, **matches**, a **copper sulfate solution** in a **beaker**, a **nail** or **screw**, a **lighter**, a **small beaker**, **beaker tongs**, a **test tube**, and a **test tube holder**. **Looking at the materials and lab we will be using, what are the safety precautions we should take to protect ourselves and materials during the investigation?**

Prep for Reaction 3

1) Take the **nail or screw** and write down how it appears now.

2) Stand it up in the small amount of **copper sulfate** solution and come back and look at it after the other reactions are done.

Reaction 1

3) Place the aluminum pan on the digital scale and zero it out.
4) Take some **steel wool** and place it in the pan on a digital scale. What is the mass of the steel wool?

5) What is the color of the steel wool before burning?

6) Take the lighter and light the steel wool. Place the lighter's flame into the steel wool and observe the reaction. What do you see happening during the reaction?

7) Notice the color change. What is the color of the burned steel wool?

8) What is the mass of the burnt steep wool?

9) Why do you think the mass changed?

10) This reaction was a combustion reaction where the iron in the steel wool combined with oxygen in the air, using fire, forming the copper oxide. Write a word equation of the reaction.

11) Balance the equation below for this chemical reaction.

$$Fe + O_2 \rightarrow Fe_2O_3$$

Reaction 2

12) Take some **baking soda** ($NaHCO_3$) and place it in a test tube; make sure you hold the test tube with the test tube holder. Take the lighter and heat the test tube with the baking soda in it. What do you see forming on the inside of the glass of the test tube?

13) Now light a match and place it inside the mouth of the test tube. What happens to the flame?

14) What color is the solid at the bottom of the test tube?

15) What you saw was water form on the inside of the test tube; and carbon dioxide gas form that snuffed the flame of the match out. Sodium carbonate is what is left. Write a word equation for the reaction.

16) Balance the chemical equation below for this reaction.

$$NaHCO_3 \rightarrow CO_2 + H_2O + Na_2CO_3$$

Reaction 3

17) Go back to your **nail or screw** and carefully pull it out of the liquid. How does it look now?

18) The iron and the copper switched places. The iron is now combined with the sulfate, and the copper is by itself. Write a word equation for the reaction.

19) Balance the equation for this reaction.

$$Fe + CuSO_4 \rightarrow Fe_2(SO_4)_3 + Cu$$

Acid-Base Reaction

20) Write a word equation for hydrochloric acid mixing with sodium hydroxide that makes water and table salt.

21) Balance the equation for this reaction.

$$HCl + NaOH \rightarrow H_2O + NaCl$$

Removing Carbon from Sugar

Directions and Questions:

Perform this reaction under a **fume hood** or go **outside**. Some unhealthy gasses and heat are produced from this reaction, so once it has started, stand back and watch but do not touch. You will need **safety goggles**, an **apron**, **sugar**, **baking soda**, **lighter fluid**, a **ceramic bowl with sand**, and a **long-necked lighter**. **Looking at the materials and lab we will be using, what are the safety precautions we should take to protect ourselves and materials during the investigation?**

1) Now take the ceramic bowl of sand and have your teacher spray some lighter fluid on it. Place a mixture of 40 g sugar and 10 g baking soda on it. Make sure there is nothing near the bowl. Have a fire extinguisher ready if the fire gets out of hand. Then with a long-neck lighter, light it on fire. What do you see happening?

2) The sugar reacts with the oxygen producing carbon dioxide and water. When the oxygen runs out, the sugar breaks down to solid carbon (what looks like a black snake) and water vapor. The baking soda changed to carbon dioxide, water, and sodium carbonate (this is what is stealing the oxygen to help the sugar create a black snake). Write the word equation for the reaction taking place.

3) Balance the reaction taking place.

$$\mathbf{NaHCO_3 + C_{12}H_{22}O_{11} + O_2 \rightarrow CO_2 + H_2O + Na_2CO_3}$$

Wait until it cools before touching it. The black snake will stain whatever touches it, so be careful as you dispose of it using your teacher's instructions.

Conservation of Mass

Directions and Questions:

You will need a **scale**, a **Ziploc bag**, **baking soda**, **vinegar**, and a **disposable pipette**. **Looking at the materials and lab we will be using, what are the safety precautions we should take to protect ourselves and materials during the investigation?**

1) Take a small Ziploc bag and put a spoonful of baking powder in it. Suck up some vinegar into the pipette and place the pipette in the bag. Seal the bag shut.
2) Find the mass of the bag and its contents before the reaction. What is the mass?

3) Squeeze the vinegar from the pipette into the baking soda. What do you observe?

4) Find the mass of the bag and its contents after the reaction. What is the mass?

5) Why do you think we made the reaction take place inside a sealed bag?

6) What did you see and feel that let you know a chemical reaction took place?

7) Was it an endothermic or exothermic reaction? How did you know?

8) Compare the mass of the bag and its contents before the reaction and after.

9) Why did the results come out as they did?

10) Use your textbook or the internet to find how the conservation of mass is stated, and write it here.

11) How did this reaction show the conservation of mass?

The Law of Conservation of Mass

Directions and Questions:

You will need **safety goggles**, an **apron**, a **graduated cylinder**, 10 mL of **potassium iodide solution** in a **250 mL Erlenmeyer flask**, a **test tube** half full with a **lead nitrate solution** placed in the Erlenmeyer flask so that it does not mix with the potassium iodide, and a **rubber stopper** sealing the flask. **Looking at the materials and lab we will be using, what are the safety precautions we should take to protect ourselves and materials during the investigation?**

1) Measure the mass of the Erlenmeyer flask and its contents. What is it?

2) Carefully and slowly invert the flask, letting the test tube's contents mix with the flask's contents while keeping the stopper sealed tight. What did you observe?

3) What did you see that indicated a chemical reaction took place?

4) Measure the mass of the flask and its contents. What is it?

5) How did this reaction show the conservation of mass?

6) Why did we seal the flask?

Conservation of Life: Photosynthesis and Respiration

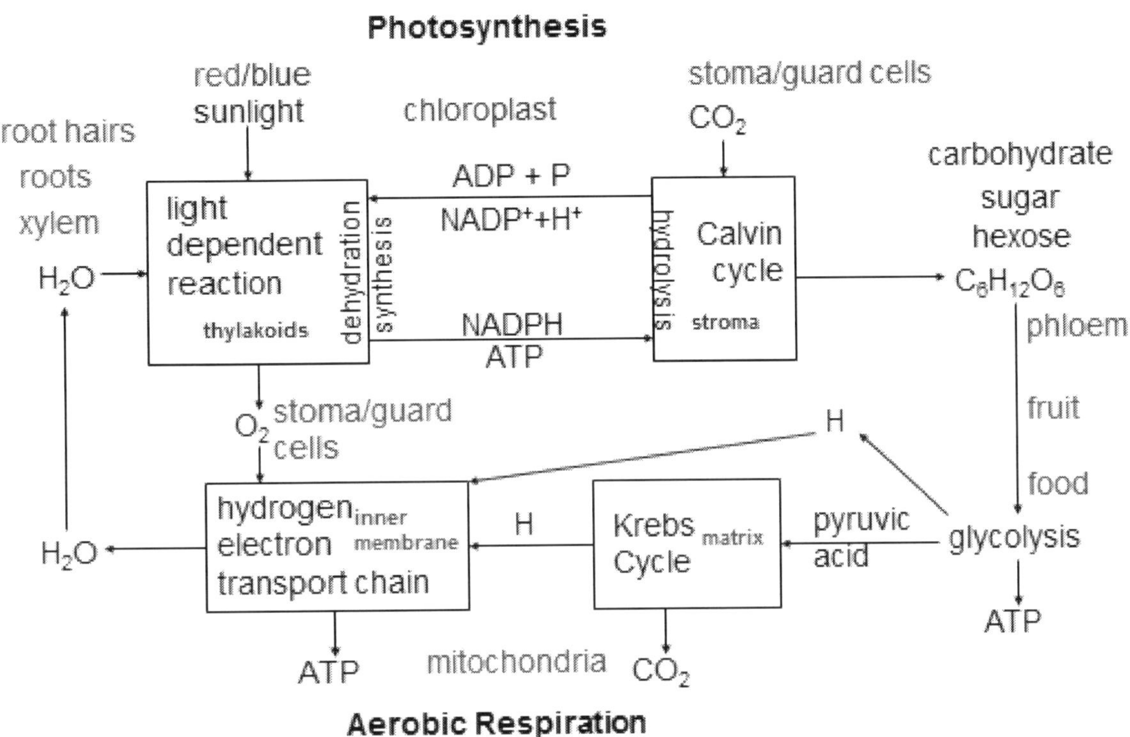

Directions:

Use the diagram above to answer the following questions.

1) Write the equation for photosynthesis and balance the reaction.

2) Write the equation for aerobic respiration and balance the reaction.

3) In both equations, trace where each element of the reactants go to make the products.

4) How are the two reactions similar to each other?

5) How does the conservation of mass, in this case, show that life has balance?

6) Plants and algae go through both photosynthesis and respiration. Animals only go through respiration. What would happen to life on Earth if we lost the plants and algae?

7) Keeping this in mind, which do you think formed first: the process of aerobic respiration or photosynthesis? Explain why.

8) How was this diagram a good model showing the conservation of mass in photosynthesis and respiration?

9) What careers would use the information shown by this activity?

Virtual Investigations that go with Conservation of Mass

ExploreLearning.com

Chemical Equations

Stoichiometry

Chemical Changes

Balancing Chemical Equations

Limiting Reactants

Moles

Cell Energy Cycle

Photosynthesis Lab

Plants and Snails

Phet.colorado.edu

Balancing Chemical Equations

Ratio and Proportion

Reactants Products and Leftovers

Physicsclassroom.com/Concept-Builders/Chemistry:

Stoichiometry Relationships

Mole Conversions

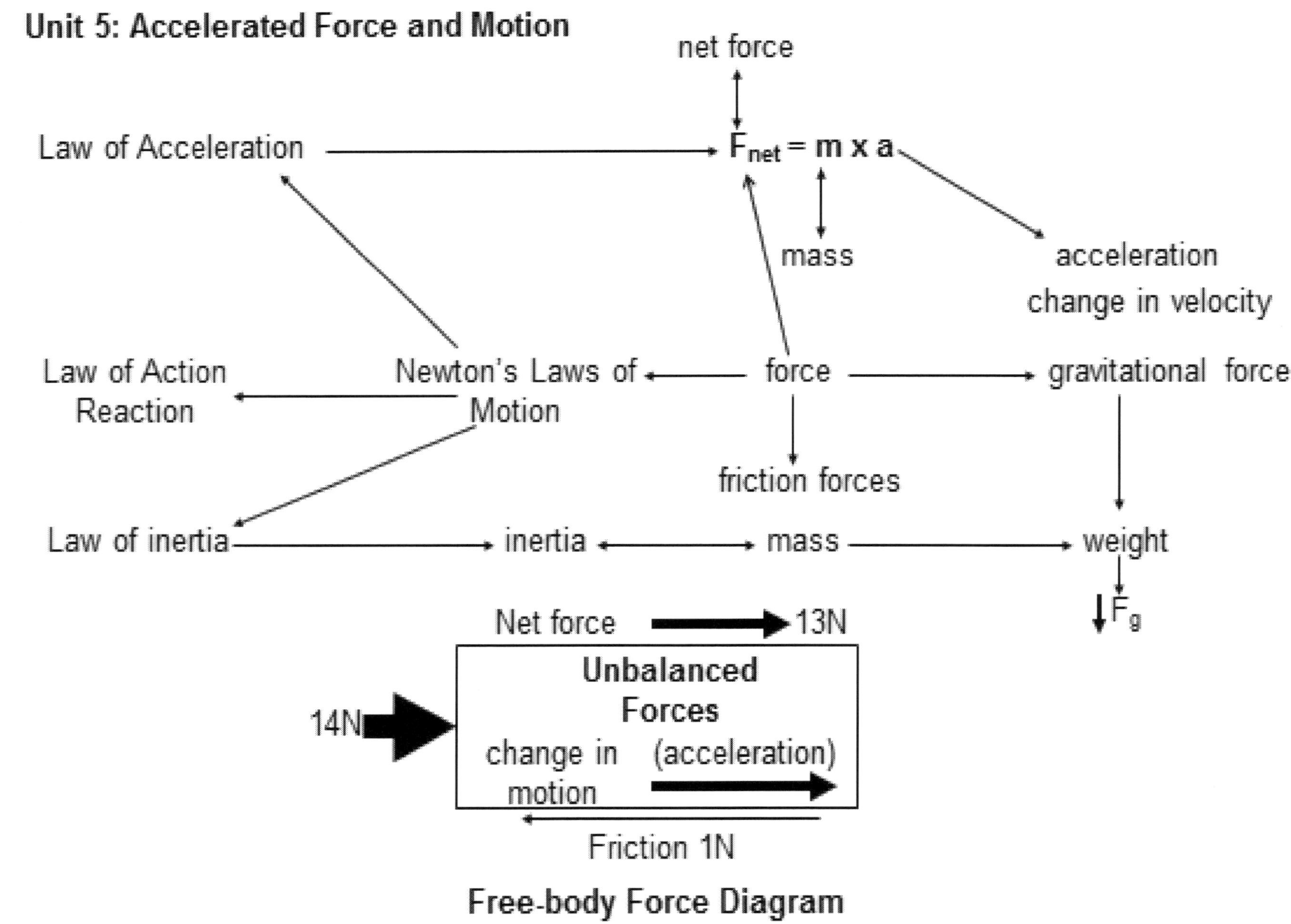
Unit 5: Accelerated Force and Motion
net force
Law of Acceleration
F_{net} = m x a
mass
acceleration
change in velocity
Law of Action Reaction
Newton's Laws of Motion
force
gravitational force
friction forces
Law of inertia
inertia
mass
weight
F_g
Net force
13N
Unbalanced Forces
14N
change in motion
(acceleration)
Friction 1N
Free-body Force Diagram

Marbles in Motion

Directions:

Get a segment of **hot wheels track**, **small stickers**, a **stopwatch**, and a **marble**. Have something that one side of the track can be placed on to raise that end to create a ramp. **Looking at the materials and lab we will be using, what are the safety precautions we should take to protect ourselves and materials during the investigation?**

1) Set the ramp so the ramp's bottom is along the edge of a tile on the floor. Most tiles in schools are 1 foot in length. Clear a path for 5 feet.
2) Adjust the height of the ramp so that the marble will just make it past 5 feet.
3) Place small stickers on the floor at the ramp's base and each foot past the ramp's base. The last one is 5 feet away from the base of the ramp.
4) Place a small sticker on the ramp to mark where you will place your marble for each trial to let it roll down the ramp; this keeps the distance your marble will be accelerating down the ramp constant.
5) Place your marble on the ramp and let it roll down (do not push). Time with a **stopwatch** how long it takes for the marble to move from the ramp's base to one foot away.
6) Repeat #5 four more times. Record the data in Data Table 1 below.
7) Repeat #s 5 and 6 for the distances 2 feet, 3 feet, 4 feet, and 5 feet away.
8) Find the average time for each distance.
9) Then calculate the average velocity for each distance by taking the distance and dividing it by the average time and write that in Data Table 1.

Data Table 1

Trial	1 foot	2 feet	3 feet	4 feet	5 feet
1					
2					
3					
4					
5					
Average Time					
Average Speed					

10) Take the average speed and plot them on the graph to make a speed-distance graph; this will look similar to a velocity-time graph since the longer the distance, the more time it takes. The shape we see will be the same as looking at accelerated motion on a velocity-time graph.

  a. This graph will be the same shape we would see for constant velocity motion on a distance-time graph.

Velocity vs. Distance Graph

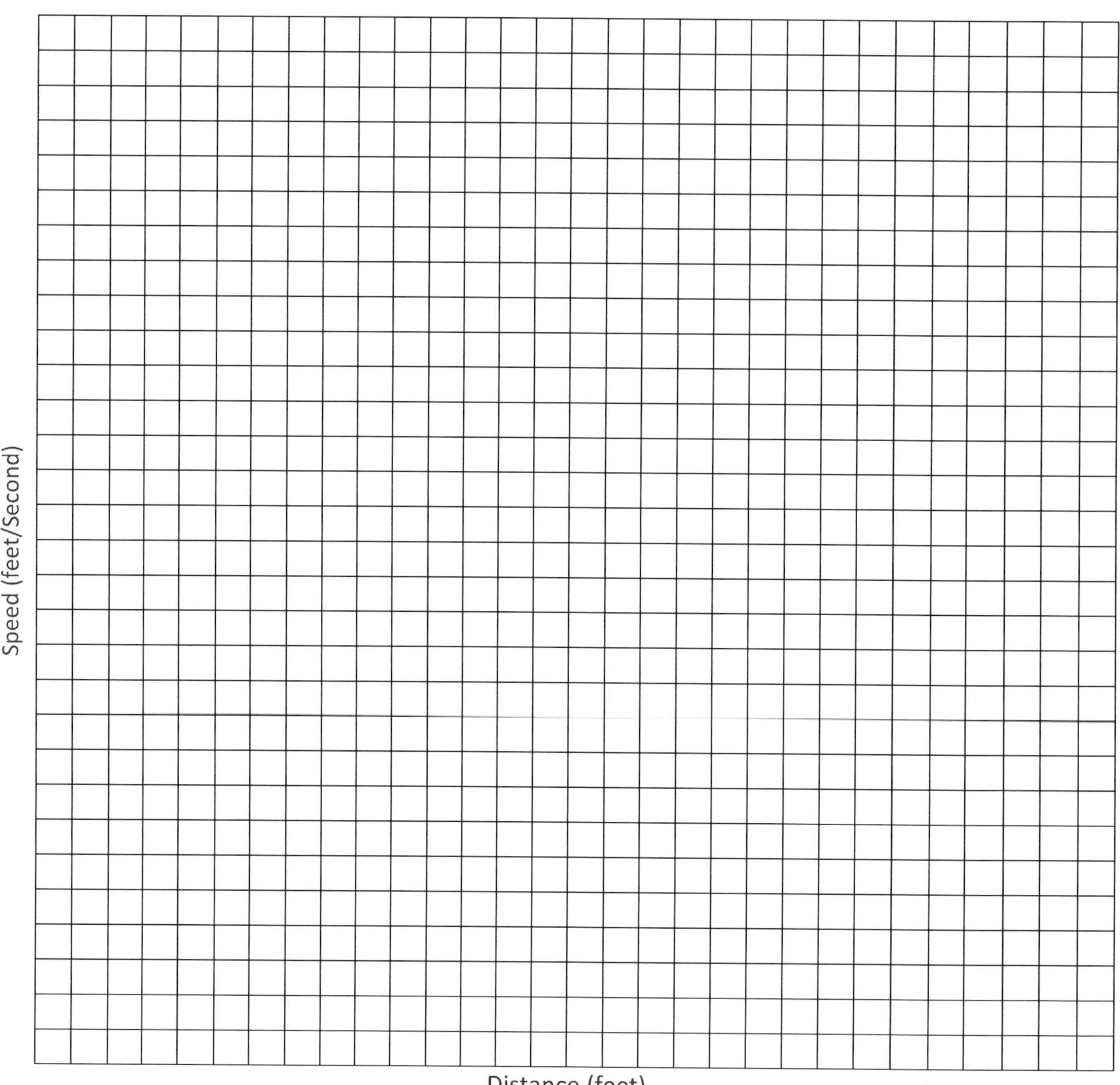

Questions:

1) Describe the motion of the marble as it moved down the ramp. If you have to, place the marble on the ramp and let it go to observe it move down the ramp.

2) What force caused the marble to speed up on the ramp?

3) How could we make the marble have a higher velocity at the bottom of the ramp?

4) Describe the motion of the marble as it moved across the floor. If you have to, place the marble on the ramp again and observe it roll across the floor.

5) What caused the marble to slow down across the floor?

6) How could we make the marble slow down faster across the floor?

7) At what distance would the average velocity be happing? At this distance would be the instantaneous velocity that is the same as the average velocity.

8) What was the shape of this graph? This shape of the graph is what acceleration looks like on a velocity-time graph.

9) What conditions would you need for the marble to have no positive or negative acceleration?

Ball Bounce

Directions:

You will need to get a **small wire basket,** a **large bouncy ball,** and attach a **motion detector** to an **interface** connected to a **computer** with **Logger Pro**. **Looking at the materials and lab we will be using, what are the safety precautions we should take to protect ourselves and materials during the investigation?**

1) Take the small wire basket and place it over the motion detector, so the ball (large bouncy ball) that you drop will not hit it. The spacing of the wires on the basket needs to be wide enough not to be detected by the motion detector.

2) When placing the basket over the motion detector, make sure the basket's gap is directly over the sensor, so the motion detector will only see the ball.

3) In Logger Pro, open folder Physics with Vernier and file #06 Ball Toss.

4) Press "Collect," drop the ball, and let it bounce on the basket. It is best if the ball bounces a few times to see what is happening in the data.

5) Look at the graphs and have the students see where each bounce is. Notice for each bounce; the graph sizes get less and less. Why would that happen?

6) Move the display to see only one bounce for all three graphs. Use that display to label the ball's motion on all three graphs simultaneously. Students can use the picture on the next page (similar to what you should see) to label what happened in the graphs they made.

7) Place a label on the graph where the ball is at the highest point, where the ball moves up, where the ball moves down, and where the ball is in contact with the basket.

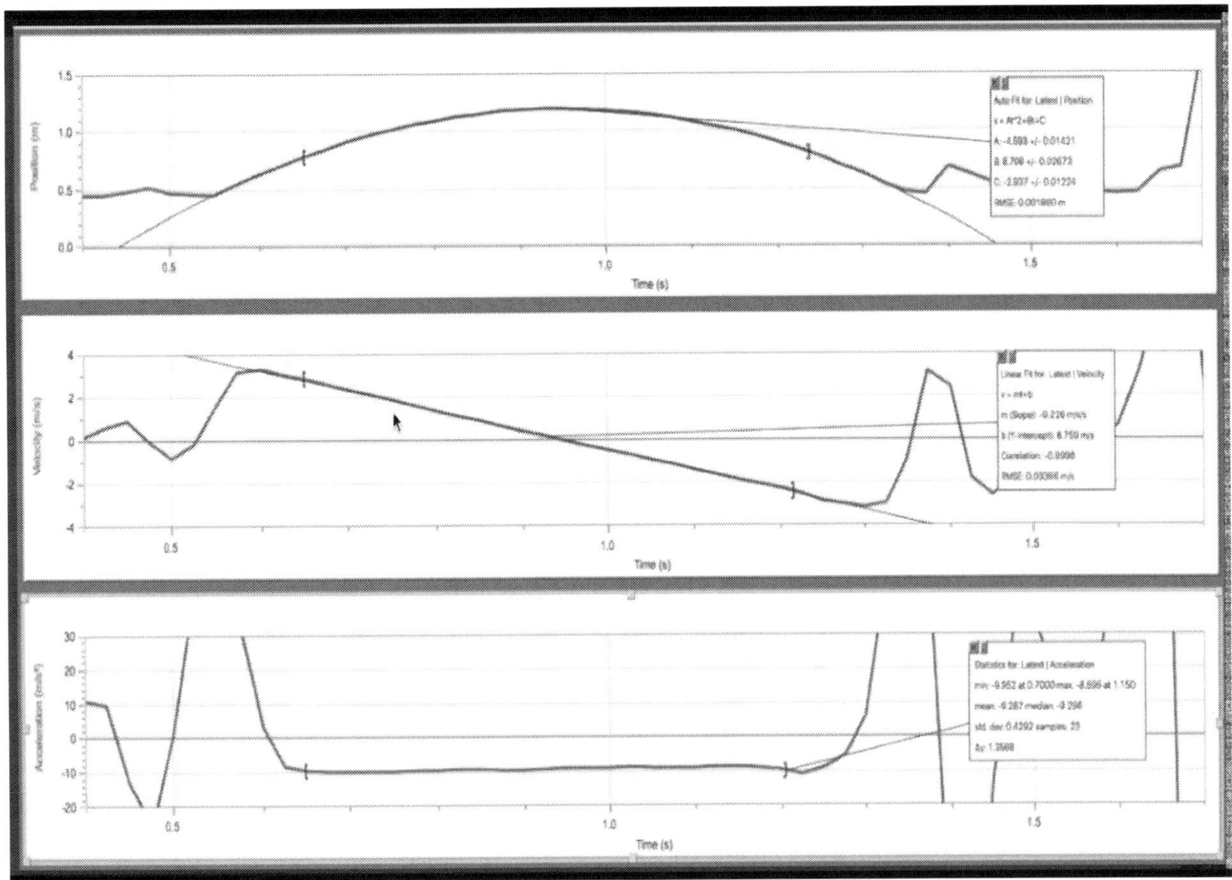

Questions:

1) What type of motion is this?

2) What is the shape of the graph of the ball as it goes through negative acceleration for:
   a. Distance vs. Time graph

   b. Velocity vs. Time graph

   c. Acceleration vs. Time graph

3) Compare this with the shape of constant velocity for:
   a. Distance vs. Time graph

b. Velocity vs. Time graph

c. Acceleration vs. Time graph

Cart on a Ramp

Directions:

You will need to get a **spring cart** with **Vernier's Dynamics System** with a **wall** on the end and a **motion detector** attached to an **interface** connected to a **computer** with **Logger Pro**. **Looking at the materials and lab we will be using, what are the safety precautions we should take to protect ourselves and materials during the investigation?**

1) Set up the spring cart on a ramp like one used in Vernier's Dynamics System, where one end of the ramp is propped up in the air.

2) Place the motion detector on the top end of the ramp with the sensor facing down the ramp.

3) In Logger Pro, open the folder Physics with Vernier and file #03 Cart on a Ramp.

4) Hold the cart at the top of the ramp with the spring side down and press "Collect," and let go of the cart, allowing it to roll down the ramp and bounce off the wall. It is best if the cart bounces a few times to see what is happening in the data.

5) Look at the graphs and have the students see where each bounce is. Notice for each bounce; the graph sizes get less and less. Why would that happen?

6) Move the display to see only one bounce for all three graphs. Use that display to label the cart's motion on all three graphs simultaneously. Students can use the picture on the next page (similar to what you should see) to label what happened in the graphs they made.

7) Place a label on the graph where the cart is at the highest point, where the cart moves up, where the cart moves down, and where the cart is in contact with the ramp's bottom wall.

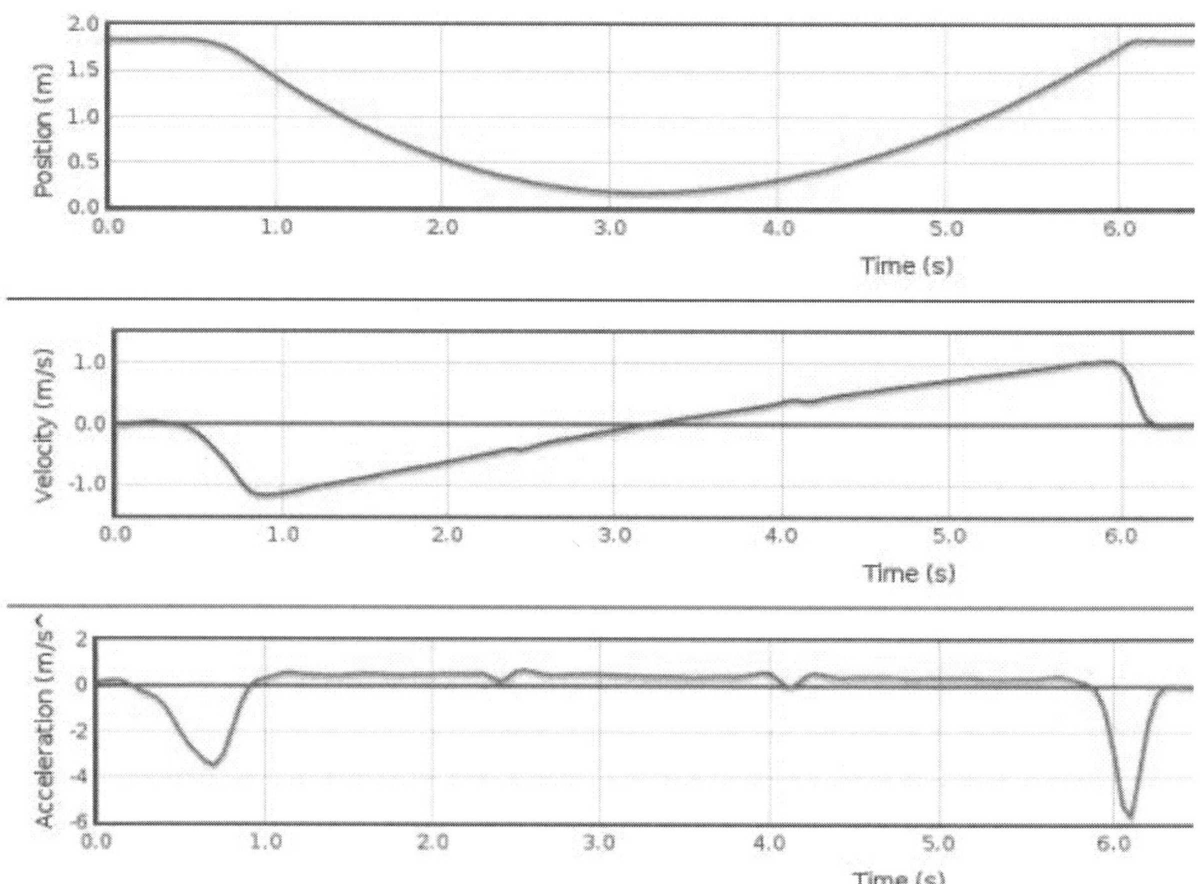

Questions:

1) What type of motion is this?

2) What is the shape of the graph of the cart as it goes through positive acceleration for:
   a. Position vs. Time graph
   b. Velocity vs. Time graph
   c. Acceleration vs. Time graph

3) Compare this with the shape of constant velocity for:
    d. Distance vs. Time graph

    e. Velocity vs. Time graph

    f. Acceleration vs. Time graph

Picket Fence Free Fall

Directions:

You will need Vernier's **Picket Fence**, a **photogate** attached to an **interface** connected to a **computer** with **Logger Pro**, and a **clamp** and **ring stand** to secure the photogate. **Looking at the materials and lab we will be using, what are the safety precautions we should take to protect ourselves and materials during the investigation?**

1) Use the clamp to fix the photogate to the ring stand and move the ring stand to the table's edge.

2) In Logger Pro, open the Physics with Vernier folder and file # 05 Picket Fence.

3) Click "Collect" to allow the photogate to turn on when you drop your picket fence through the photogate.

4) Then hold your picket fence over the photogate letting it hang down. Drop the photogate so that the alternating black and clear bands flow through the photogate. If one part of the picket fence does not go through in the drop, your reading will be off.

5) Look at your data, and make sure your slope for the velocity vs. time graph is a straight line; if it isn't, repeat #s 3 & 4.

6) If the graph's slope is straight, click the linear fit button and record the line's slope in the top data table on the next page; this is the acceleration of the picket fence dropping through the photogate.

7) Repeat steps 3-6 four more times.

8) Find the Minimum and Maximum values of the trials, calculate the average by adding up all five slopes, and then dividing by 5. Write this value in the middle data table on the next page. It should be close to 9.81 m/s^2.

9) Find the precision by taking the lowest number of your average and 9.81 and dividing by the other, then multiply by 100, giving you the % accuracy. Write that down in the bottom data table.

Data Tables

Trial	1	2	3	4	5
Slope (m/s^2)					

	Maximum	Minimum	Average
Acceleration (m/s^2)			

Acceleration due to gravity, g	9.81 m/s^2
Precision	%

Questions:

1) How does the force of gravity affect the motion of a falling object?

2) How does this relate to Newton's 2nd Law of Motion?

3) What is the shape of the position vs. time graph for each trial?

4) What is the shape of the velocity-time graph for each trial?

5) How close is your average acceleration compared to the 9.81 m/s^2?

Measuring the Effects of Air Resistance

Directions:

You will need two **stopwatches** and two **pieces of paper**. **Looking at the materials and lab we will be using, what are the safety precautions we should take to protect ourselves and materials during the investigation?**

1) Take one piece of paper and wad it up into a ball. Leave the other paper flat.
2) Hold both pieces of paper at the same height and drop them simultaneously. Make sure the flat paper is horizontal to the floor when you drop it. Have one person time the flat paper and the other person time the paper wadded up in a ball.
3) Write the times in Data Table 1 below.
4) Repeat the procedure in #s 2-3 two more times.
5) Find the averages of the times by adding the three times up and dividing by 3. Put this data in Data Table 1 below.

Data Table 1

Type of Paper	Trial 1 Time (s)	Trial 2 Time (s)	Trial 3 Time (s)	Average Time (s)
Flat				
Wadded				

Questions:

1) Which one accelerated faster?
2) Which one reached its terminal velocity first?
3) Which part of the investigation showed Newton's 1st Law of Motion?

4) Which part of the investigations showed Newton's 2nd Law of Motion?

5) Which part of the investigation showed Newton's 3rd Law of Motion?

6) How does this relate to why we use parachutes when jumping out of planes?

7) Which careers need to know this information to protect people and equipment?

Elevator Lab

Directions and Questions:

Take a **1 kg mass** into an **elevator** with a **digital scale**. **Looking at the materials and lab we will be using, what are the safety precautions we should be taking to protect ourselves and materials during the investigation?**

1) Place the 1 kg mass on the digital scale on the elevator floor.
2) What is the mass in the scale say before the Elevator moves?
3) Press a button in the elevator to move the elevator down. How does the reading on the scale change when the elevator starts to move down?

   (gets higher, gets lower)

4) How does the reading on the scale change when the elevator starts to slow down?

   (gets higher, gets lower)

5) Press a button in the elevator to move the elevator up. How does the reading on the scale change when the elevator starts to move up?

   (gets higher, gets lower)

6) How does the reading on the scale change when the elevator starts to slow down?

   (gets higher, gets lower)

7) When did the mass seem to have less weight?
8) Try to explain why.

9) When did the mass seem to have more weight?

10) Try to explain why.

11) How did this investigation show Newton's 2nd Law of Motion?

12) Where in this investigation did we see Newton's 1st Law of Motion?

13) Where in this investigation did we see Newton's 3rd Law of Motion?

Newton's Relay Race

Directions and Questions:

You will need a **broom**, a **bowling ball**, a **basketball**, and a kid's **rubber ball**. **Looking at the materials and lab we will be using, what are the safety precautions we should take to protect ourselves and materials during the investigation?**

Accelerating an object from rest: we will be observing inertia – resistance to change motion (Newton's 1st Law)

1) Place the bowling ball on the floor. Push the ball with a broom in a sweeping motion to cause the bowling ball to accelerate. When does the bowling ball accelerate?

2) How easy was it to accelerate?

3) When does it move at a constant speed?

4) Place the kid's rubber ball on the floor and push this ball the same way you did with the bowling ball. When does the ball accelerate?

5) How easy was it to accelerate?

6) When does the ball move at a constant speed?

7) Which ball had the most inertia?

8) Draw a force diagram of the action of accelerating the ball.

Stopping an object: observing inertia (Newton's 1st Law) and momentum – resistance to stop motion

9) Get the bowling ball moving, then stop it with the broom. Do the same with the kid's rubber ball. Which ball was harder to stop?

10) When sitting still, which object has the most inertia?

11) Draw a force diagram of the action of stopping a moving ball.

Turning an object 180 degrees: observing inertia (Newton's 1st Law) and momentum

12) Get the bowling ball and start moving it, then stop it and turn it 180 degrees. Do the same for the kid's rubber ball. Which ball was easier to change direction?

13) Which ball had the most inertia?

Applying a constant force on the ball: observing force and acceleration (Newton's 2nd Law)

14) Go to a long hallway and make sure it is clear. Get the bowling ball moving by pushing it with the broom; see what happens if you try to put a constant force on the ball. Do the same for the kid's rubber ball. What happened to the speed of both balls?

15) Could you keep doing it?

16) Which ball could you apply the force for the longest amount of time? Why?

17) Which ball accelerated the fastest?

18) How could you make the balls move at a constant velocity?

Relay Race: observing inertia (Newton's 1st and 2nd Laws)

19) Make a course that changes direction several times (I have made my student go around the center demo table in my room) to push the different balls around in a race. Divide up into three equal teams, each with a broom. One team will push a bowling ball with a broom, one team will push a basketball with a broom, and one group will push the kid's rubber ball with the broom. Predict which team will finish the course first with all its team members.

20) Have the students do a relay race to move the balls through the course (be very careful of the bowling ball and make sure kids know to move out of its way if it comes at them). Which team won?

21) Why were they able to win?

22) Which team came in last? Why?

23) When was Newton's 3rd Law of Motion happening during today's investigation?

Newton's Second Law

Directions:

You will need a **scale**, a **cart**, a long **rubber band**, a **dual-range force sensor**, and **low g accelerometer**, and an **interface** connected to a **computer** with **Logger Pro**. **Looking at the materials and lab we will be using, what are the safety precautions we should take to protect ourselves and materials during the investigation?**

1) Stack the dual-range force sensor on top of the cart and the accelerometer on the force sensor. Have them face the same direction as the wheels will be moving, so the force and acceleration will be measured in the same direction. Tightly wrap the rubber band around the system, holding it all together.

2) Connect the dual-range force sensor to channel 1 on the interface. Connect the low g accelerometer to channel 2.

3) Open Physics with Vernier folder and file # 09 Newton's Second Law.

4) Click "Collect" to collect data. Holding the hook of the dual-range force sensor, roll the cart back and forth in the direction the wheels move. Vary the forces, both small and large.

5) What is the shape of the force vs. time graph?

6) What is the shape of the acceleration vs. time graph?

7) Click the examination button. Move the mouse across one of the graphs. When the force is at its maximum, what is the acceleration? (maximum or minimum)

8) Click on the Force vs. Acceleration graph and click the linear fit button. Record the slope of this line in Data Table 1.

9) Find the mass of the cart and sensors. Write this in Data Table 1.
10) Add/fix a 500 g mass to the cart, repeat the procedure for #4 and 8, and record the slope of the force vs. acceleration graph in Data Table 1.

11) Find the mass of everything in the cart now. Write that in Data Table 1.

Data Table 1

Cart	Slope of graph	Mass (kg)
Cart and Sensors		Kg
Cart, sensors, and 500 g		Kg

Questions:

1) How are the Force vs. Time and Acceleration vs. Time graphs similar for the two trials?

2) How are they different?

3) Compare the slope of the Force vs. Acceleration graph and the mass. What does the slope represent?

4) Write a general formula for the three variables: force, mass, and acceleration.

5) What is the unit for the slope of the Force vs. Acceleration graph?

6) What is the relationship between force and acceleration in the equation?

7) What is the relationship between the mass and acceleration in the equation?

8) How could this information help you run away from a rhinoceros chasing you?

9) If you have a bowling ball and a baseball, each is suspended by a separate rope and hit each with a baseball bat, which ball will have the biggest change in motion? Explain why.

Fan Cart Lab

Directions:

You will need a **scale**, a **mass**, a **cart**, a **fan** to fix to the cart, a **motion detector** attached to an **interface** connected to a **computer** with **Logger Pro**, and a **Dynamics System** from Vernier. **Looking at the materials and lab we will be using, what are the safety precautions we should take to protect ourselves and materials during the investigation?**

1) Use the scale to measure the mass of the cart. Write this in Data Table 1.
2) Position the motion detector at one end of the track. Position the cart in front of the motion detector so it will move away from the motion sensor.
3) Start the fan. Click the "Collect" button. Allow the cart to accelerate down the track. Click "Stop." Grab the cart before it falls off the track.
4) Highlight the line on the Velocity vs. time graph. Click the "Fit" button and choose linear fit. The slope of the line is the acceleration of the cart. Record the acceleration in Data Table 1.
5) Repeat #3-4 two more times and put the accelerations in Data Table 1.
6) Calculate the average acceleration by adding the three values and dividing by 3.
7) Add a mass to the fan cart. Record the amount of the mass in the data below.
8) Repeat the procedures in #3-6 with the added mass. Put this data in Data Table 1.

Data:

Mass of the cart and fan __________kg Amount of mass added __________kg

Data Table 1

Cart	Mass (kg)	Trial 1 Acceleration	Trial 2 Acceleration	Trial 3 Acceleration	Average Acceleration
Cart + Fan					
Cart + Fan + Mass					

Questions:

1) Using the formula F = m x a, calculate the force of the fan on the cart. What is that force?

2) Do the same for the cart with the mass on it. What is that force?

3) Compare the answers for #s 1 and 2.

4) Why did they have different accelerations then?

5) What may be a source of error not figured in the equation?

Water Bottle Rockets

Directions and Questions:

You will need a **water bottle rocket launcher**, a **2-liter bottle**, and an **air pump** with a **pressure gauge. Looking at the materials and lab we will be using, what are the safety precautions we should take to protect ourselves and materials during the investigation?**

1) Fill the water bottle half full with water.
2) Angle the launcher straight up at a 90-degree angle to the ground (gives the rocket its highest distance to travel in the air).
3) Pump 20 pounds of pressure into the rocket. Start a stopwatch when you launch the rocket and stop it when it reaches the highest point in the air. What was the time?

4) What do you think will happen to the time in the air and the launch's height if we double the pressure?

5) Pump 40 pounds of pressure into the rocket. Start a stopwatch when you launch the rocket and stop it when it reaches the highest point in the air. What is the time?

6) How did changing the pressure affect the flight of your rocket?

7) If you were to launch again, what other variables could we change to affect the rocket's height?

8) How do you think changing the rocket's mass will affect the force of gravity?

9) How would it affect the inertia?

10) How is Newton's 1st Law of Motion affect the rocket launch?

11) How does Newton's 2nd Law of Motion affect the rocket launch?

12) Draw a force diagram to show the variables affected in Newton's 2nd Law of Motion.

13) How is Newton's 3rd Law of Motion seen in the launch?

14) Draw a force diagram to show how the launch shows Newton's 3rd Law of Motion.

Virtual Investigations that go with Accelerated Force and Motion

ExploreLearning.com:

Free Fall Tower Gizmo

Sled Wars Gizmo

Feed the Monkey

Free-Fall Laboratory

Fan Cart Physics

Force and Fan Carts

Gravity Pitch

Orbital Motion – Kepler's Laws

Golf Range

PhET.colorado.edu:

Maze Game

Motion 2D

The Moving Man

Ladybug Motion

Energy Skate Park

Energy Skate Park: Basics

Friction

Gravity and Orbits

Projectile Motion

Physicsclassroom.com:

Physics Interactives:

Newton's Laws of Motion

Force

Balanced and Unbalanced Forces

Free-Body Diagram

Rocket Sledder

Which One Doesn't Belong?

Skydiving

Elevator Ride

Atwood's Machine

Kinematics

Graphs & Ramps

Circular and Satellite Motion

Uniform Circular Motion

Race Track

Roller Coaster Model

Roller Coaster Design

The Elevator Ride

Orbital Motion

The value of g

Gravitation

Concept Builders:

Newton's Laws of Motion

Balanced and Unbalanced Forces

Force and Motion

Change of State

Recognizing Forces

Match That Free-Body Diagram

Which One Doesn't Belong? Card Sort

Newton's Second Law – Equations as Guides to Thinking

Net Force (and Acceleration) Ranking Tasks

Air Resistance and Skydiving

F<u>net</u>=m*a

Solve It! with Newton's Second Law

Kinematics

Acceleration

Name That Motion

Motion Diagrams

Graph That Motion

Match That Graph

Velocity – Time Graphs

Dots and Graphs

Words and Graphs

Which One Doesn't Belong

Circular and Satellite Motion

Circular Logic

Case Studies – Circular Motion

Force and Free-Body Diagrams in Circular Motion

Universal Gravitation

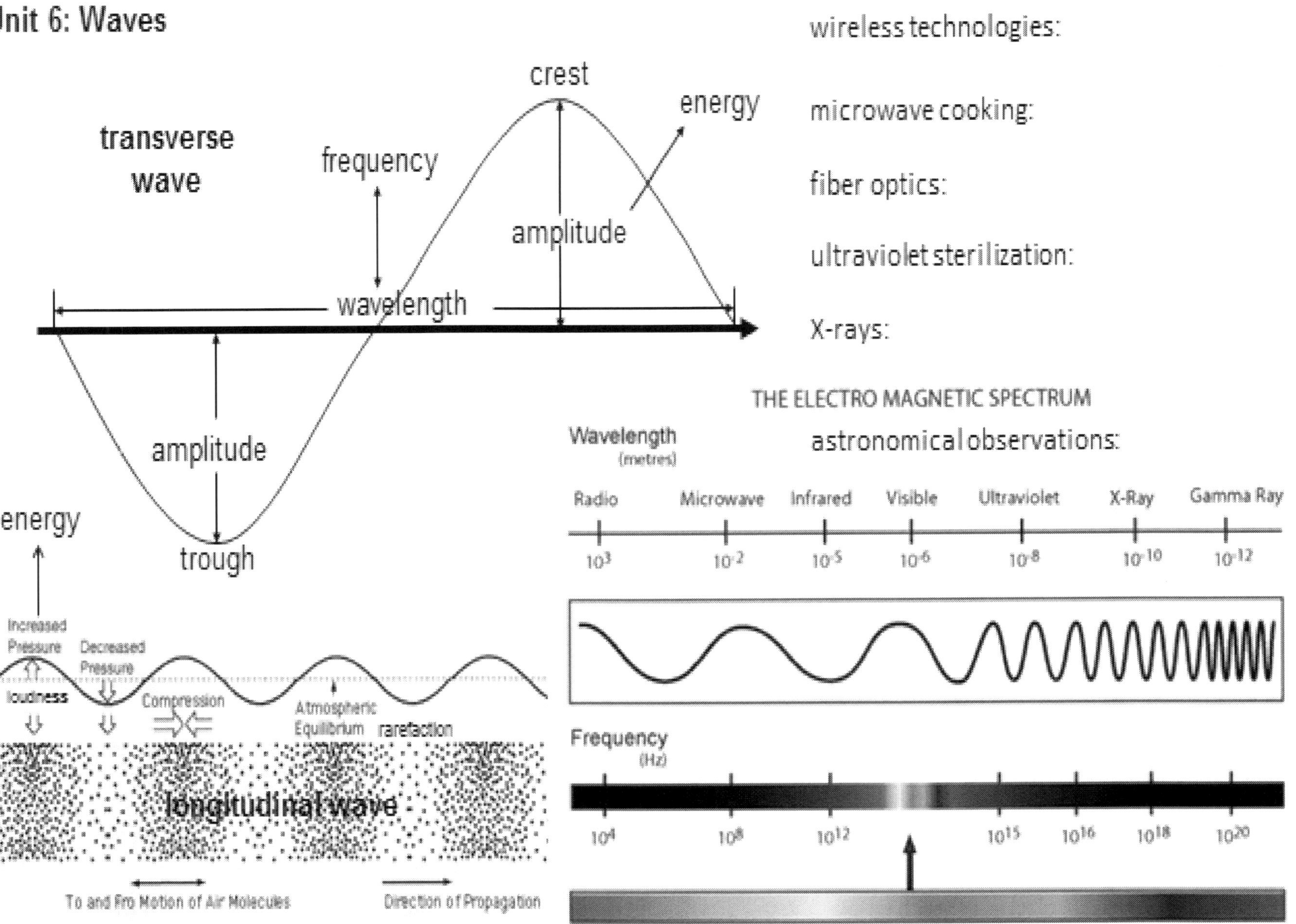
Unit 6: Waves
transverse wave
frequency
crest
energy
amplitude
wavelength
amplitude
energy
trough
wireless technologies:
microwave cooking:
fiber optics:
ultraviolet sterilization:
X-rays:
THE ELECTRO MAGNETIC SPECTRUM
astronomical observations:
Wavelength (metres)
Radio
Microwave
Infrared
Visible
Ultraviolet
X-Ray
Gamma Ray
10^{3}
10^{-2}
10^{-5}
10^{-6}
10^{-8}
10^{-10}
10^{-12}
Frequency (Hz)
10^{4}
10^{8}
10^{12}
10^{15}
10^{16}
10^{18}
10^{20}
Increased Pressure
Decreased Pressure
loudness
Compression
Atmospheric Equilibrium
rarefaction
longitudinal wave
To and Fro Motion of Air Molecules
Direction of Propagation

Measuring Wave Properties

Directions:

You will need a long springy **telephone cord**, a **stopwatch**, and a **meter stick**. **Looking at the materials and lab we will be using, what are the safety precautions we should take to protect ourselves and materials during the investigation?**

1) Stretch your phone cord and measure how long it is when you will make waves with it.
2) Have two people, one on each end, hold the phone cord. Have one person send a pulse down the cord by having them move their hand back and forth once. Have a third person measure how long it takes for the wave to make it to the other person. In Data Table 1 on the next page, write this data for all three wave times (since all the waves will travel at the same speed down the same medium).
3) Oscillate the cord back and forth to create a standing wave that is half a wavelength long; this will make it look like one hump is in the wave. Make sure you keep the rhythm. In your data table, multiply your cord's length by 2 to get the wavelength for wave one. Write this in Data Table 1.
4) Have the person with the stopwatch; once they see the rhythm on the cord, say start, and they will start the stopwatch and measure 10 seconds and then say stop when the ten seconds are done. The person moving their hand will count how many times they move their hand to the right during the 10 seconds. Write this data down in Data Table 1 on the next page as the wave count.
5) This time, oscillate the cord back and forth to create a standing wave that is one wavelength long; this will look like there are two humps. The wavelength for wave 2 is the length of the cord.
6) Repeat the procedure for #4 and write the data for wave 2 in Data Table 1.
7) This time oscillate the cord back and forth to create a standing wave with one and a half wavelengths long; this will look like three humps. The wavelength for wave 3 is 2/3 the length of the cord.
8) Repeat the procedure for #4 and write the data for wave 3 in Data Table 1.
9) Calculate the wave speed by taking the length of the cord and dividing it by the wave time. Write this in Data Table 1 on the next page for all three waves.
10) Calculate the frequency by taking the wave count and dividing it by 10 seconds. Write this in Data Table 1 on the next page for all three waves.

Data Table 1

Wave	Cord Length (m)	Wave Time (s)	Wave Speed (m/s)	Wave Count	Wavelength (m)	Frequency (Hz)
1						
2						
3						

Questions:

1) Why was the wave speed the same for all three waves?

2) What is the relationship between wavelength and frequency?

3) The cord has a natural frequency and what you created were three octaves of that frequency. Where is the word octaves used that has to do with waves?

4) How does frequency relate to pitch?

5) So if someone sings at a high pitch, is the wavelength long or short? Explain why.

6) If someone sings at a low pitch, is the wavelength long or short? Explain why.

Observing Waves in a Slinky

Directions:

You will need a standard to long **slinky** to make both compression waves and transverse waves. You will do this with one person holding one end of the slinky and another person holding the other. One person will move one end; the other will hold still. Each person needs to take a turn at moving the slinky to make the waves described below. You will show this to your teacher. **Looking at the material and lab we will be using, what are the safety precautions we should take to protect ourselves and materials during the investigation?**

1) Make a compression wave with high frequency by moving the end of your slinky forward and backward in the same plane as the slinky quickly.

2) Make a compression wave with low frequency by moving the end of the slinky forward and backward in the same plane as the slinky but slowly.

3) Make a compression wave with high amplitude by repeating the procedure in #2 but making bigger, more violent pushes down the slinky.

4) Make a compression wave with low amplitude the same as the procedure in #2 but making smaller, less violent pushes down the slinky.

5) Make a transverse wave with high frequency by quickly moving the slinky's end perpendicular to the slinky.

6) Make a low-frequency transverse wave by repeating the procedure in #5 but move the slinky more slowly.

7) Make a transverse wave with a high amplitude by repeating the procedure in #5 but moving the end of the slinky a bigger distance perpendicular to the slinky.

8) Make a transverse wave with low amplitude by repeating the procedure in #5 but not as big a distance as #7.

9) Now make a compression wave with high frequency and low amplitude.

10) Now make a compression wave with low frequency and high amplitude.

11) Now make a transverse wave with high frequency and high amplitude.

12) Now make a transverse wave with low frequency and low amplitude.

Observing Sound

Directions:

You will need a **wire hanger** and **string.** Tie two pieces of string on either side of the wire hanger. **Looking at the materials and lab we will be using, what are the safety precautions we should take to protect ourselves and materials during the investigation?**

1) Wrap the string around each of your index fingers and clank the hanger against your desk. How does the hanger sound?

2) Put your fingers in your ears and clank the wire hanger against the desk. How does the hanger sound now?

Questions:

1) Do sounds travel better through the air or the string?

2) Which medium do you think sound travels the fastest through? Put the types of medium (the solid, liquid, and gas) in order from fastest to slowest.

3) What if there is no medium for sound to travel through, will there be any sound?

4) When the Death Star blows up in Star Wars, can that really make a sound in outer space? Explain why?

Coffee Can Phones

Directions and Questions:

Take two **coffee cans** and poke a small hole in the bottom of each of them. Cut a long piece of **string** that reaches across the room, put the ends through each can, and tie a knot in them, fixing them to both cans. **Looking at the materials and lab we will be using, what are the safety precautions we should take to protect ourselves and materials during the investigation?**

1) Pull the string tight while holding the cans and talk through them. Can you be heard in the other can? Why do you think that is?

2) Have someone pinch the string with their fingers halfway across. Can you be heard in the other can now? Why do you think that is?

3) Let the string loosen and droop. Talk again. Can you be heard in the other can? Why do you think that is?

4) Combine with a group next to you, cross your strings, and have them touch while the strings are taught. Have someone talk into a can. Who can hear in their cans?

5) Have someone pinch with their fingers where the strings cross. Can anyone hear now?

6) Explain how sound travels from one can to the other.

Music Test Tubes

Directions:

You will need four **test tubes**, a **test tube holder**, and different **water** amounts placed in each test tube. Test tube one, leave empty. Test tube 2 fill it ¼ full of water. Test tube 3 fill 1/3 with water. Test tube 4 fill with ½ full with water. **Looking at the materials and lab we will be using, what are the safety precautions we should take to protect ourselves and materials during the investigation?**

1) Predict how the toned will sound different when you blow across the top of the test tube. Rank the predicted sounds from 4 being the lowest to 1 being the highest. Write that in Data Table 1 below.
2) Blow across test tube one until you get a tone produced. Do the same for each of the test tubes. Rank the order from the 4 being the lowest tone to 1 being the highest.
3) The tone is created by how large/long the tube is for the air to vibrate. The longer the tube, the longer the wavelength. Empty and wash the test tubes when you are done.

Data Table 1

Test Tube	Amount of Water	Predicted Tone Differences	Tone Difference
1	empty		
2	¼ full		
3	1/3 full		
4	½ full		

Questions:

1) Describe how the tones changed depending on the amount of water in the test tube.

2) How did the pitch depend on the height of the water?

3) Why are the tones different from the different test tubes?

4) Explain how resonance amplifies the sound of a test tube.

5) How do the natural frequencies of the columns of air in each test tube differ?

6) Compare how the test tubes make music with how a flute makes music.

7) How is the flute different from the test tubes?

8) How does the music industry use the physics of waves?

Singing Glasses and the Dancing Toothpick

Directions and Questions Part 1:

You will need a **flat toothpick**, **crystal glasses of different sizes**, some **crystal glasses of the same size** filled with different **water** amounts, and the **internet**. **Looking at the materials we will be using, what are the safety precautions we should take to protect ourselves and materials during the investigation?**

1) Wet your finger and rub it around the rim of one of the glasses until you hear a hum.
2) Repeat the procedure in #1 for different sizes of glasses. Which glasses (larger or smaller) have a deeper tone or pitch?

3) Which glasses have a higher tone or pitch?

4) Fill some of the glasses of the same size with different amounts of water. Which glass has the lowest pitch?

5) Which glass has the highest pitch?

6) Research on the internet why your results came out the way they did. What causes the sounds to change when you change the glass's size or how much water is in the glass?

Directions and Questions Part 2:

1) Now take two of the glasses you used in Part 1 that are the same size and place them near each other. Balance a flat toothpick on the rim of one of the glasses.
2) Rub the rim of the other glass to make it hum. What do you notice happens to the toothpick?

3) Discuss with your class and teacher, then explain why this happens.

4) What do natural frequency and resonance have to do with this phenomenon?

Playing the Rubber Band

Directions:

You will need a **plastic tub** and three different **rubber bands** of the same length with different widths. **Looking at the materials and lab we will be using, what are the safety precautions we should take to protect ourselves and materials during the investigation?**

1) Find the mass of each of the rubber bands. Write that in Data Table 1 below.
2) Put the rubber bands around the tub. Try to have the same tension on each rubber band.
3) Pluck each rubber band and tell how they differ. Rank the pitch from 3 being the lowest to 1 being the highest. Put that in Data Table 1 below.

Data Table 1

Rubber Band	Mass	Pitch
Skinny		
Medium		
Thick		

Questions:

1) How does the width of the rubber band affect the pitch?

2) The rubber band that takes the longest to move back and forth will have the lowest frequency/longest wavelength. Try to explain why this happens. (Remember the Law of Inertia.)

3) How do you think that affects the acceleration of the rubber bands as they move back and forth?

4) How do you think the length of the rubber band would affect the pitch?

5) How do string thickness and length affect how string instruments sound and are played?

Music has Patterns

Part 1 Directions:

You will need a **digital keyboard** and a **microphone probe** attached to an **interface** connected to a **computer** with **Logger Pro**. **Looking at the materials and lab we will be using, what are the safety precautions we should take to protect ourselves and materials during the investigation?**

1) In Logger Pro, open the Physics with Vernier folder and file #35, Mathematics of Music.
2) Position the microphone near the opening where the sound comes out of the instrument. Press "Collect." Play a middle C for the first note. Play it until you see the wave form on the screen. Record the frequency in Hz. Write this in Data Table 1.
3) Now play the next higher note to repeat the procedure in #2. Repeat this until you have recorded all the notes' frequency going up in the scale until you reach the next C.
4) Now calculate the ratio of the first note to middle C; this is done by taking the current note's frequency and dividing it by the middle C frequency.
5) Take the decimal from the ratio and try to find the fraction that is closest to that decimal, and write your answer like the example shown in Data Table 1. Compare this to the different sizes of wrenches we use in a toolbox or on a ruler in proportions to an inch; you will see a pattern. Any variation off of the pattern shows the instrument is out of tune.

Data Table 1

Key	Note	Frequency (Hz)	Ratio to C	Ratio to C Fraction
1	C			1 and 0
2	D			
3	E			1 & ¼
4	F			
5	G			

6	A			1 & 2/3
7	B			
8	C			

Questions Part 1:

1) What is the frequency ratio to middle C with the next C higher?

2) How long is the wavelength with this C compared to the middle C? Hint: use the wave formula.

3) Are there any other notes that we did not play? What do you think they are?

Part 2 Directions:

1) To see why they are there, play all the keys in order from middle C (even the black keys). Write the frequencies in Hz down in Data Table 2.
2) Then find the ratio to the previous note by taking the current note and dividing it by the frequency of the previous one. What number did you get for each?

Data Table 2

Key	Note	Frequency	Ratio to the prev. note
1 White	C		
2 Black	C sharp		
3 White	D		
4 Black	E flat		
5 White	E		
6 White	F		

7 Black	F sharp		
8 White	G		
9 Black	A flat		
10 White	A		
11 Black	B flat		
12 White	B		
13 White	C		

Questions Part 2:

1) What number did you get for each ratio to the previous note?

2) Besides the pattern of frequency ratios, what other patterns do we see in music?

3) If there are no patterns in the sound, what do we call it?

4) What should you have in sound to be able to call it music?

The Doppler Effect

Directions and Questions:

You will need a toy **football that whistles** when you throw it. Stand between two people that will throw it back and forth over your head. **Looking at the materials and lab we will be using, what are the safety precautions we should take to protect ourselves and materials during the investigation?**

1) When the ball is flying over your head, pay attention to the whistle's pitch. How was the sound before the ball reached you compared to the sound after it passed you?

2) What happens to the pitch of the whistle as it passes over your head?

3) Why do you think this happened?

4) When would the ball move into the waves, shortening the wavelength coming to your ear as it flies in the air?

5) When would the ball move away from the waves, lengthening the wavelength coming into your ear as it flies in the air?

6) How does this explain why a siren sounds higher as it approaches and changes to a lower pitch as it passes?

Making a Rainbow

Directions and Questions:

You will need a **sunny day** and a **hose** or **spray bottle** to make mist. **Looking at the materials and lab we will be using, what are the safety precautions we should take to protect ourselves and materials during the investigation?**

1) Make mist with a hose or a spray bottle with the sun behind you until you see a rainbow. Try to see the entire rainbow. What is the shape of the rainbow?

2) What is the order of the colors of the rainbow?

3) Which color is on the outside?

4) Which color is on the inside?

5) Which color do you think has the longest wavelength? Explain why.

6) Which color do you think has the shortest wavelength? Explain why.

7) AM radio can be heard across the country because the wavelength is longer and can get past the curvature of the Earth; this is not the case for FM radio; once out of town, the radio station seems to go out. Which color do you think would be able to go around the particles in the air when the light passes through the thickest part of the atmosphere? Explain why.

8) What color is the sun when we see it on the horizon (sunrise or sunset)? Explain why.

9) What color is the moon when we see it on the horizon? Explain why.

Extention: If you have some **diffraction gradient glasses**, put those on and explain what you see happen to all the light you see. These act similar to how the water diffracts the light we just saw through the rainbow.

Polarization of Light

Directions:

You will need a **light source**, two **polarizing filters**, and a **light sensor** attached to an **interface** connected to a **computer** with **Logger Pro. Looking at the materials and lab we will be using, what are the safety precautions we should take to protect ourselves and materials during the investigation?**

1) In Logger Pro, open the folder Physics with Vernier and the file #28 Polarization of Light. Press "Collect."
2) Turn on your light source and turn off all the other lights in the room. Place your light sensor at a distance away from your light source and measure the light intensity. Write the data in Data Table 1 below.
3) Place a polarizing filter between the sensor and the light. Measure the light intensity and write the data in Data Table 1 below.
4) Place another polarizing filter between the sensor and the light in a way that still allows light to pass through. Measure the light intensity and write the data in Data Table 1 below.
5) Rotate one filter 90° to block the light from the light source going to the sensor. Measure the light intensity now and write that data in Data Table 1 below.

Data Table 1

Setup	Light Intensity
Just light	
Light through a filter	
Light through 2 filters	
Light through 2 filters at 90°	

Questions:

1) How did the light intensity change when the filter was placed in front of the light?

2) How did the light intensity change when two filters were placed in front of the light?

3) What happened to the light intensity when one lens was rotated 90°?

4) Hold the two filters up to the light and rotate one. What do you see?

5) Why do you think this happens?

6) Polarized sunglasses do the same thing. How can you tell if a pair of sunglasses are really polarized?

3D Glasses

Directions and Questions:

You will need two pairs of **3D glasses. Looking at the materials and lab we will be using, what are the safety precautions we should take to protect ourselves and materials during the investigation?**

1) Face the glasses towards each other directly in front of each other. What color do you see through the lenses when looking through both sets of lenses simultaneously?

2) Move the glasses closest to you back and forth from left to right. How do the colors change when looking through both sets of lenses on the glasses simultaneously?

3) Turn one set of glasses perpendicular to the ground. How did the color change as you looked through both sets of lenses?

4) Face the glasses in the same direction (looking at them from behind) and turn one pair perpendicular to the ground; how does the lens color change when looking through both sets of lenses of the glasses at the same time?

5) Put one pair of glasses on and look at the other with both eyes open. How do the lenses look on the other pair of glasses?

6) Close your right eye; what do you see now in the other glasses?

7) Close your left eye; what do you see now in the other glasses?

8) Polarized sunglasses block the horizontal light from hitting your eyes and allow the vertical light through; this keeps you from seeing glare from the sun reflecting off surfaces. How could you tell if lenses at the sunglasses store are really polarized?

9) The slits on the 3D glasses have one lens with microscopic slits that block light coming at you vertically, whereas the other lens blocks the light coming at you horizontally. If they are turned at right angles to each other, it blocks all light. Two projectors show the same movie, just staggered with different types of light so that your right eye sees one projector, and your left eye sees the other; this allows you to see a 3D image on the screen.

Light Pipes

Directions and Questions:

You will need **fiber optics** to observe some **light sources of different colors. Looking at the materials and lab we will be using, what are the safety precautions we should take to protect ourselves and materials during the investigation?**

1) Shine a light on the side of the fiber optics. Can you see the light coming out the ends?

2) Shine light into the end of a fiber optic wire. Do you see the light coming out the other end?

    a. Can you see the light coming out the sides?

3) Keep shining the light through one end of the fiber optic and bend the fiber optic. Can you still see the light coming out the other end?

    a. Can you see it coming out the sides?

4) Try shining the light on the other end. Does light still come out at the opposite end?

5) Why do you think they call these light pipes?

6) How can they be used to send information from one computer to another to have the computers talk to each other?

7) Who do you think benefits from this technology and information?

8) Try shining a different color of light down one end of the pipe. What do you see?

Water Refraction

Directions:

You will need a **square tank** half-filled with **water**, an **Erlenmeyer flask** filled with **water**, and a **ruler**. **Looking at the materials and lab we will be using, what are the safety precautions we should take to protect ourselves and materials during the investigation?**

1) Put the ruler behind the tank with it up and down perpendicular to the floor. Look through the water from the front at the same level as the tank. Draw what you see below.

2) Look at the ruler through the tank from the front above the tank. Draw what you see below.

3) Turn the tank so that a corner is facing you. Put the ruler on the right-hand side; where do you see the ruler? Draw it below.

Questions:

1) Why do you think the images appear different through the same tank of water at different angles?

2) How would the image you see be different if the water's surface reflected light like a mirror instead of bending it?

3) How would the image change if the container was curved, bending out?

4) Pull out an Erlenmeyer flask filled with water. Look through it, and tell how things look different at different distances.

5) How does this explain how convex lenses work?

Test Tube Lenses

Directions:

Fill a **glass test tube** with **water** and seal it with a **rubber stopper**. Keep your finger or thumb over the rubber stopper so that the stopper does not fall off and spill water. **Looking at the materials and lab we will be using, what are the safety precautions we should take to protect ourselves and materials during the investigation?**

1) Set the test tube on the paper over the title of this lab. Write your observation in Data Table 1 about what you see through the test tube.
2) Hold the test tube approximately 1 cm over the title of this lab and observe it again. Record your observations in Data Table 1 below.
3) Repeat this three more times, increasing height approximately a centimeter each time. Write your observations in Data Table 1 below.

Data Table 1

Height	Observation of **Test Tube Lenses**
Right on the surface	
1 cm above the surface	
2 cm above the surface	
3 cm above the surface	
4 cm above the surface	

Questions:

1) Are the images you see real or virtual?

2) How high above the surface did the image become inverted?

3) What kind of lens does the test tube make (concave or convex)?

4) How does a magnifying glass compare to the test tube you looked through?

Reflection Lab

Directions:

You will need a **flat mirror**, **paper**, and a **protractor**. **Looking at the materials and lab we will be using, what are the safety precautions we should take to protect ourselves and materials during the investigation?**

1) Have your mirror facing you sitting on this paper where it says Place Mirror Here where three angles are drawn on it going into the mirror and a normal.
2) Ensure the normal is straight in and out of the mirror to properly see each of those lines pointing at you in the mirror; draw extensions for each line as if it came out of the mirror.
3) Measure the angle of incidence of the lines already there with respect to the normal.
4) Now measure the angles of reflection.

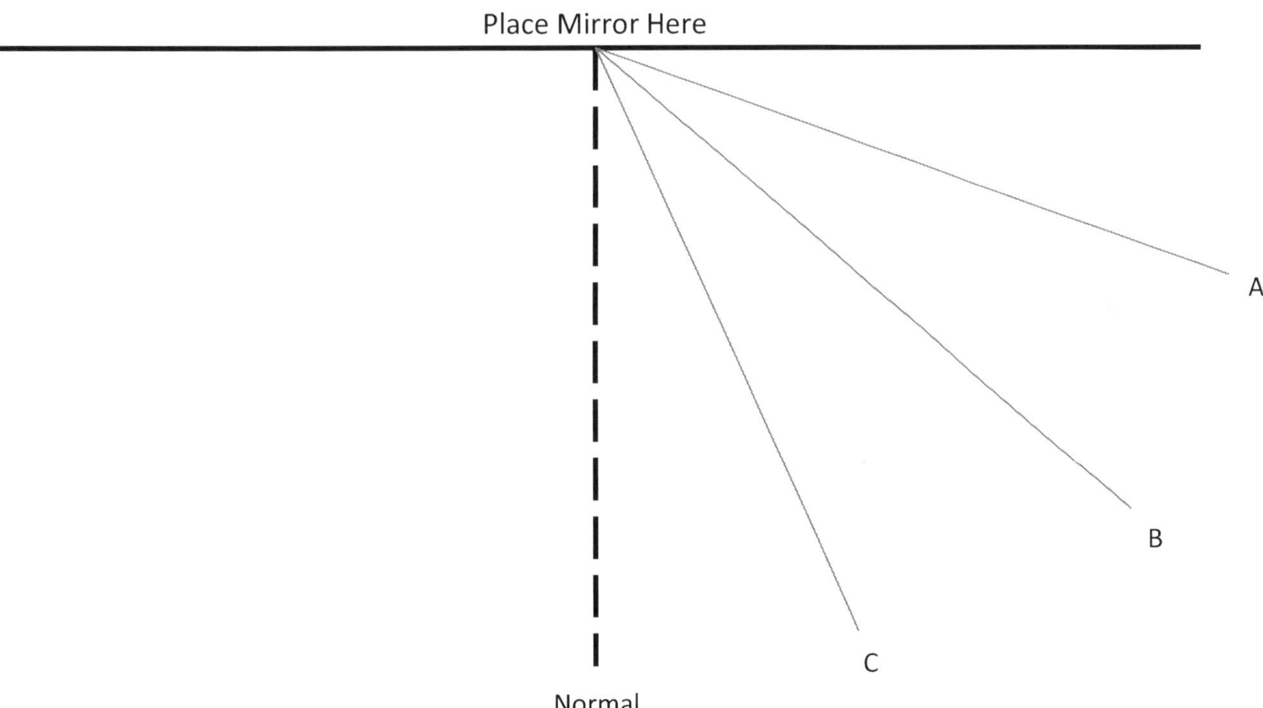

Questions:

1) What is the angle of incidence of ray A?

2) What is the angle of reflection of ray A?

3) What is the angle of incidence of ray B?

4) What is the angle of reflection of ray B?

5) What is the angle of incidence of ray C?

6) What is the angle of reflection of ray C?

7) What do you notice about the angles of the ray of reflections you drew compared to the angles of incidence?

8) Write a rule that states what you observed.

9) How would this information help you play a game of pool?

10) What is the line called where the angle of incidence and the angle of reflection is zero?

Magnifying Power

Directions:

You will need a **direct light source (bright lamp**, **flashlight**, or **sun)**, two different **power (thicknesses) lenses/magnifying glass**, and a **metric ruler**. **Looking at the materials and lab we will be using, what are the safety precautions we should take to protect ourselves and materials during the investigation?**

1) To find the focal length of both lenses, you will want to have the light source behind/above you and any flat surface in front of you. Move your lens forward and backward until you find the smallest diameter of a light spot on the front surface in front of you. Measure the distance your lens is from that surface; this is your focal length. Write this down in Data Table 1.
2) Calculate the magnifying power using focal length in cm with the equation: MP = 25cm/FL where MP = Magnifying Power, FL= Focal Length. Write this number in Data Table 1 for magnifying power using focal length.
3) To find the magnifying power using the length of an image you see, draw a 1 cm line on this paper, then place the lens over the image and see it clearly and larger. Hold your metric ruler up just above the image so you can see both the ruler and the line. Measure how long the line appears in the lens in centimeters; this now tells you the magnifying power. Place the magnifying power using the length of the image for both lenses in Data Table 1.

Data Table 1

Lens	Focal Length (cm)	Magnifying Power using Focal Length	Magnifying Power using Length of Image
Thick	(cm)		
Thin	(cm)		

Questions:

1) Describe the image formed by the magnifying glass/lens.

2) Is the image you see through the lens a real or virtual image? Tell why.

3) Compare the methods of calculating Magnifying Power. How close are the results?

4) Which method do you think is most accurate? Tell why.

5) What is happening to the light as it passes through the lens to cause the image to change?

6) How does this relate to the curvature of the lens?

7) In the movie *A Bug's Life*, an ant uses a drop of water as a lens. Is it possible to use a drop of water as a lens?

Brightness and Distance

Directions:

You will need a **light source,** a **laser light**, a **meter stick**, a **light sensor** attached to an **interface** connected to a **computer** with **Logger Pro**, or a **Vernier Dynamics System and Optics Expansion Kit**. **Looking at the materials and lab we will be using, what are the safety precautions we should take to protect ourselves and materials during the investigation?**

1) In Logger Pro, open the folder Physics with Vernier and file #29 Light Brightness Distance. Press "Collect."
2) Turn on your light source and turn off the other lights in the room. Place the 0 of the meter stick at the light source. Then place the front of the light sensor at each distance in Data Table 1 below. Measure the light intensity for each of those distances. Press "Keep" when the intensity value stabilizes at each distance. Write this data in Data Table 1 below, and press "Stop" when you have finished collecting all of the data.

Data Table 1

Distance (cm)	Intensity
5	
10	
15	
20	
25	
30	
35	
40	
45	
50	

Questions:

1) In Logger Pro, look at the graph of light intensity vs. distance. What is the relationship between distance and intensity?

2) Why do you think this happens?

3) How does this relationship help us find the distance between galaxies and stars?

4) How does this relationship relate to planets, and how close they can be to stars and possibly support life (the goldilocks zone)?

5) If the orbit of the Earth is not stable, when would temperatures be higher than normal?

    a. When would it be lower than normal?

6) Lasers are focused light made by mirrors and lenses. Try to shine a laser light into the light sensor. How does that compare to the other readings?

7) Change the distance the sensor is from the laser. Did the intensity change? Why do you think that is?

Uses of the Electromagnetic Spectrum

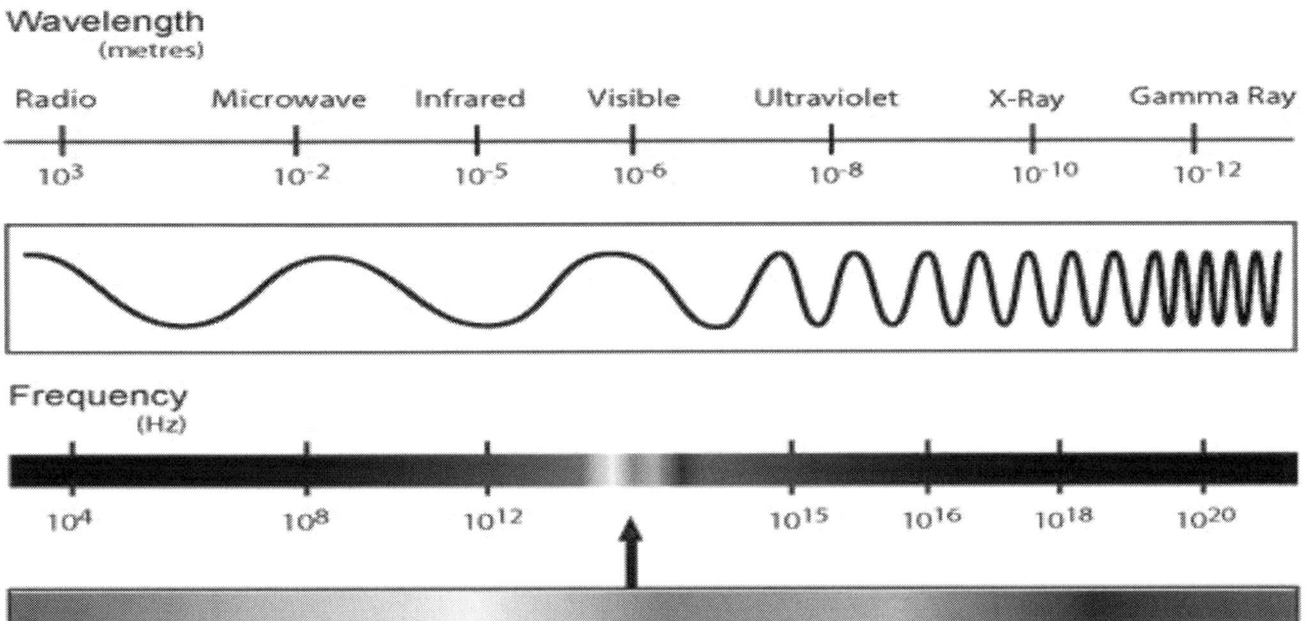

Directions and Questions:

Use the Electromagnetic Spectrum above and the **internet** to answer the questions below.

1) Electromagnetic waves have higher energy with shorter wavelengths and lower energy with longer wavelengths. Which waves above would have the highest energy, also being the most dangerous?

2) Which waves above have the lowest energy and are the most harmless?

3) Where is visible light on this spectrum?

4) Why do you think cell phones use microwaves to communicate?

5) Longer wavelengths can travel around corners easier. Which would travel farther, radio waves or microwaves?

6) Which would you see more of radio towers or cell phone towers? Explain why.

7) Why do you think remote controls use infrared waves to control TVs, drones, and remote control cars?

8) How do you think a pilot in Arizona controls a drone in the Middle East?

9) Why do you think the doctor puts a lead shield over you when you get an x-ray?

10) What do you think will give you a more detailed image, a DVD using red laser light or a DVD using blue? Explain why.

11) When using a telescope to look into outer space, which will give you more information and details, using the light waves or x-rays? Explain why.

12) Since the big Bang occurred 13.77 billion years ago, and its energy is lost over time, which category of waves would you expect to find its echo? Explain why.

13) Why do you think hospitals and restaurants use ultraviolet rays for sterilization?

14) Why do you think x-rays can see our bones?

How we use Microwave Ovens

Directions:

You will need a **microwave oven**, two **microwave bowls**, **sand**, **water**, **oven mitts**, and a **temperature probe** connected to an **interface** that is connected to a **computer** with **Logger Pro**. Microwaves cook food by flipping water molecules so fast that the friction creates heat. **Looking at the materials and lab we will be using, what are the safety precautions we should take to protect ourselves and materials during this investigation?**

1) Microwaves cook food by flipping water molecules so fast (about 2.45 billion times per second) that the friction creates heat. Pour dry sand into a microwave bowl and check the temperature of the sand with the temperature probe. Write this data in Data Table 1.
2) Place the dry sand into the microwave and run the microwave for one minute. Using oven mitts, take the bowl of sand out of the microwave and measure the temperature with the probe. Write this data in Data Table 1.
3) Subtract the two numbers and write down the change in temperature in Data Table 1.
4) Place some more sand into another bowl and add some water. Measure the temperature of the sand and water before putting it in the microwave. Write this data in Data Table 1.
5) Put the bowl in the microwave and run it for one minute. Using your oven mitts, take the bowl of sand out of the microwave and measure the temperature with the digital thermometer. Write this data in Data Table 1.
6) Subtract the two numbers and write down the change in temperature in Data Table 1.

Data Table 1

	Temperature before Microwave (^{o}C)	Temp after Microwave (^{o}C)	Change in Temperature
Dry Sand			
Wet Sand			

Questions:

1) Which bowl had a greater temperature change?

2) Why do you think this happened?

3) How does this explain why dried food stuck to the inside of the microwave does not get hot?

4) Will a microwave always cook food? Explain.

Virtual Investigations that go with Waves

ExploreLearning.com:

Waves Gizmo

Ripple Tank Gizmo

Phases Array Gizmo

Sound Beats and Sine Waves Gizmo

Longitudinal Waves Gizmo

Hearing Frequency and Volume Gizmo

Doppler Shift Gizmo

Doppler Shift Advanced Gizmo

Big Bang Theory – Hubble's Law

Earthquakes 1 Recording Station Gizmo

Earthquakes 2 Determination of Epicenter Gizmo

Photoelectric Effect Gizmo

Star Spectra Gizmo

Herschel Experiment Gizmo

Bohr Model: Introduction

Bohr Model of Hydrogen Atom Gizmo

PhET.colorado.edu:

Fourier: Making Waves

Normal Modes

Sound

Wave Interference

Wave on a String

Waves Intro

Band Structure

Bending Light

Blackbody Spectrum

Color Vision

Davisson-Germer: Electron Diffraction

Fourier: Making Waves

Geometric Optics

Lasers

Microwaves

Molecules and Light

Neon Lights and Other Discharge Lamps

Optical Quantum Control

Photoelectric Effect

Quantum Wave interference

Radiating Charge

Radio waves and Electromagnetic Fields

Simplified MRI

Physicsclassroom.com:

Physics Interactives:

Waves and Sound

Vibrating Mass on a String

Slinky Lab

Simple Wave Simulator

Wave Addition

Standing Wave Maker

Beat Patterns

Light and Color

Electromagnetic Spectrum Infographic

RGB Addition

Paint with CMY

Color Shadows

Filtering Away

Colored Filters

Stage Lighting

Viewed in Another Light

Young's Experiment

Concept Builders:

Waves and Sound

Wave Basics

Wavelength

Waves – Case Studies

Rocking the Boat

Wave Interference

Decibel Scale

Name That Harmonic: Strings

Name That Harmonic: Open-End Air Columns

Name That Harmonic: Closed-End Air Columns

Light and Color

Spectrum

Light Intensity

Color Addition and Subtraction

If This. Then That: Color Subtraction

Color Pigments

Color Filters

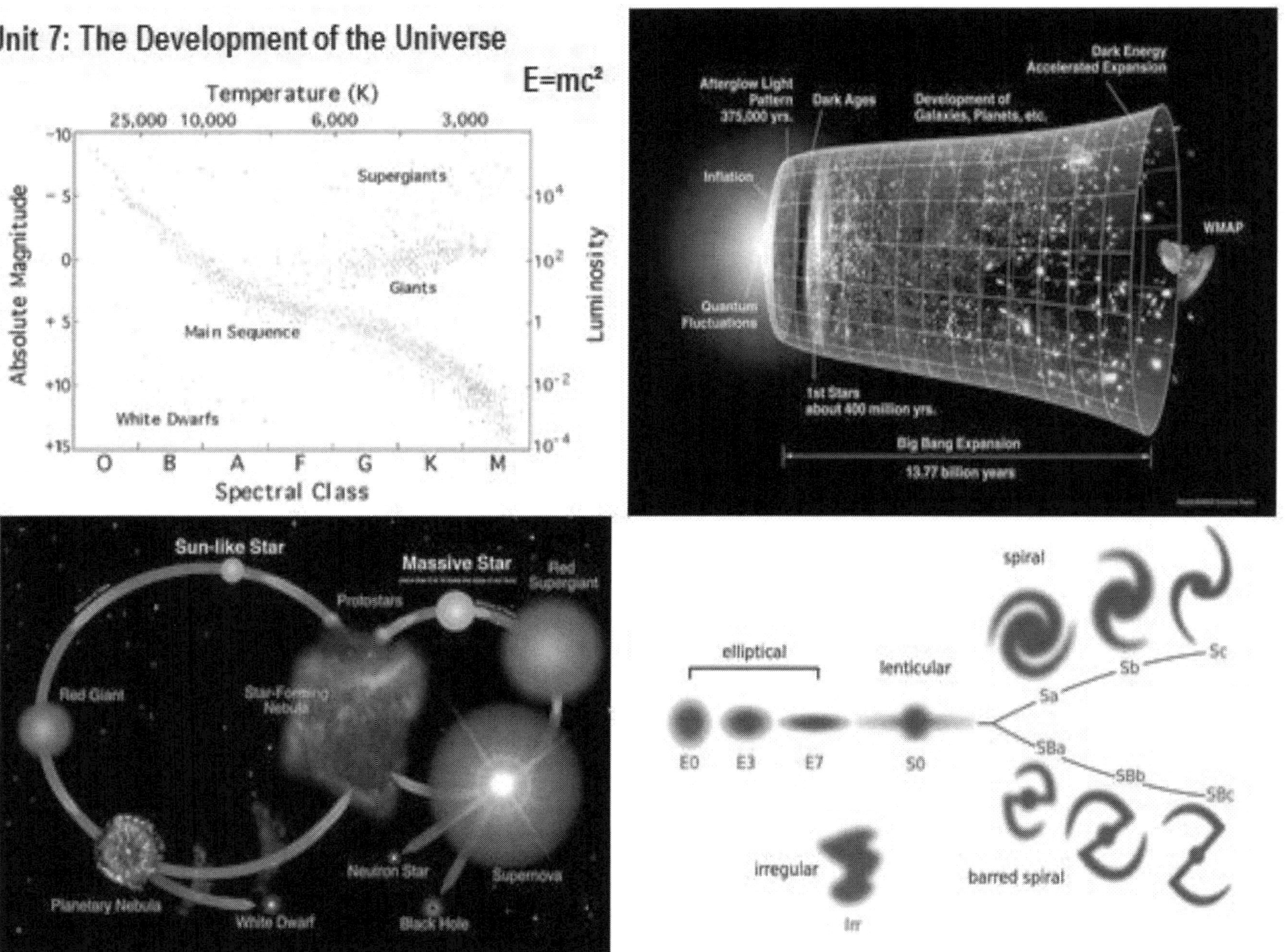
Unit 7: The Development of the Universe
E=mc²
Temperature (K)
25,000 10,000 6,000 3,000
Absolute Magnitude
-10
- 5
0
+ 5
+10
+15
Luminosity
10⁴
10²
1
10⁻²
10⁻⁴
Supergiants
Giants
Main Sequence
White Dwarfs
O B A F G K M
Spectral Class
Afterglow Light Pattern 375,000 yrs.
Dark Ages
Development of Galaxies, Planets, etc.
Dark Energy Accelerated Expansion
Inflation
Quantum Fluctuations
1st Stars about 400 million yrs.
Big Bang Expansion
13.77 billion years
WMAP
Sun-like Star
Massive Star
Red Supergiant
Protostars
Red Giant
Star-Forming Nebula
Planetary Nebula
White Dwarf
Neutron Star
Black Hole
Supernova
elliptical
lenticular
spiral
barred spiral
irregular
E0 E3 E7 S0
Sa Sb Sc
SBa SBb SBc
Irr

Star's Life Cycle

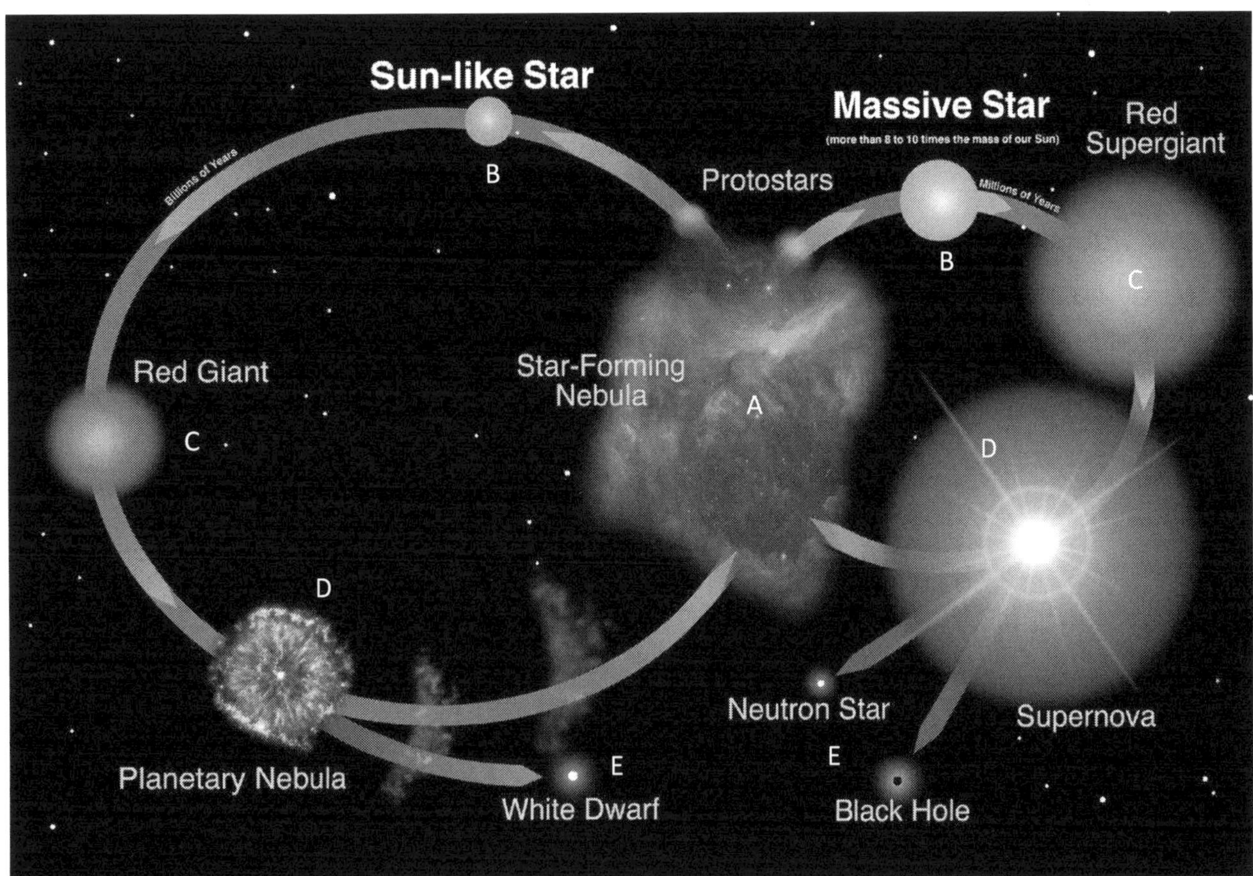

Directions:

Use the diagram above from NASA to answer questions about the life cycle of stars.

1) What is the difference in how long a star will live between normal stars like our sun and massive stars?

2) How do all stars start their life?

3) How do all stars end their life?

4) Why is this called a cycle?

5) In a star, gravity battles with fission and fusion reactions that allow it to remain a star. Fission and fusion allow it to expand, and gravity makes it collapse. As it ages, it speeds up its fuel burning and starts expanding. What letter on the diagram above are the stars expanding?

6) Then the star runs out of fuel, and its gravity makes it collapse on itself, causing a great explosion. What letter on the diagram shows the star exploding?

7) After it explodes, what will our sun leave behind?

8) After it explodes, what can massive stars leave behind?

9) How do you think an aging sun will affect life on Earth?

10) Do we have to worry about that in our lifetime? Explain.

11) Explain how a large star could give rise to smaller stars.

12) If we want to look for more evolutionarily advanced life, what type of stars should we focus on to look for planets with that kind of life? Explain.

Star Life Cycle Model

Directions:

You will need **round balloons of different sizes**, one for each group. **Looking at the materials and lab we will be using, what are the safety precautions we should take to protect ourselves and materials during this investigation?**

1) The empty balloon is like the shapeless **nebulae of gas** that will form a star.
2) Take a balloon and blow air in it until you get it to start filling up; this will represent when the **star forms**.
3) Look at the different sizes of balloons in the class; this represents the different sizes of stars.
4) As the star ages, it will speed up the burning of fuel, and the star expands. Blow air into the balloon showing the **aging of the star**.
5) Eventually, the star will run out of fuel and **explode into a supernova**. Keep blowing air into the balloon until it explodes.
6) As the stars explode, listen to the amplitude of each of the balloons; the bigger the balloon, the bigger the sound; just like the bigger the star, the bigger the explosion of that star.

Questions:

1) How did this model correctly show the lifecycle of stars?

2) What did this model not show about the lifecycle of a star? What was missing?

Our Bright Morning Star: the Sun

Directions:

Use the **internet** to go to the URL address: www.solarsystem.nasa.gov/solar-system/sun/ to help you answer the questions that follow.

1) What are the ten things we need to know about the sun?
   a.

   b.

   c.

   d.

   e.

   f.

   g.

   h.

i.

j.

2) What is the diameter of the sun?

   a. How does that compare to Earth?

3) What does the sun orbit?

4) When and how did the sun form?

5) Describe the structure of the sun.

6) Describe the surface of the sun.

7) Describe the six key features of the sun shown in the pictures.

   a. Sun Spots:

   b. Coronal Holes:

   c. Solar Flares:

   d. CME:

   e. Solar Prominence:

   f. Spicules:

Nuclear Fission and Fusion in a Star

Directions and Questions:

Use the **internet** and your **textbook** to research and explain how nuclear fission and fusion happen inside a star.

1) Draw a diagram of nuclear fission taking place in the sun.

2) What are the element isotopes involved?

3) How does the reaction get started (what goes into the reaction)?

4) What comes out of the fission reaction?

5) How does this lead to a chain reaction?

6) Draw a diagram of nuclear fusion taking place in the sun.

7) What are the element isotopes involved?

8) How is fusion different than fission?

9) What goes into the fusion reaction?

10) What comes out of the fusion reaction?

11) How do these reactions fight gravity in the sun?

12) What happens to the star if these reactions stop?

Compare and Classify Stars

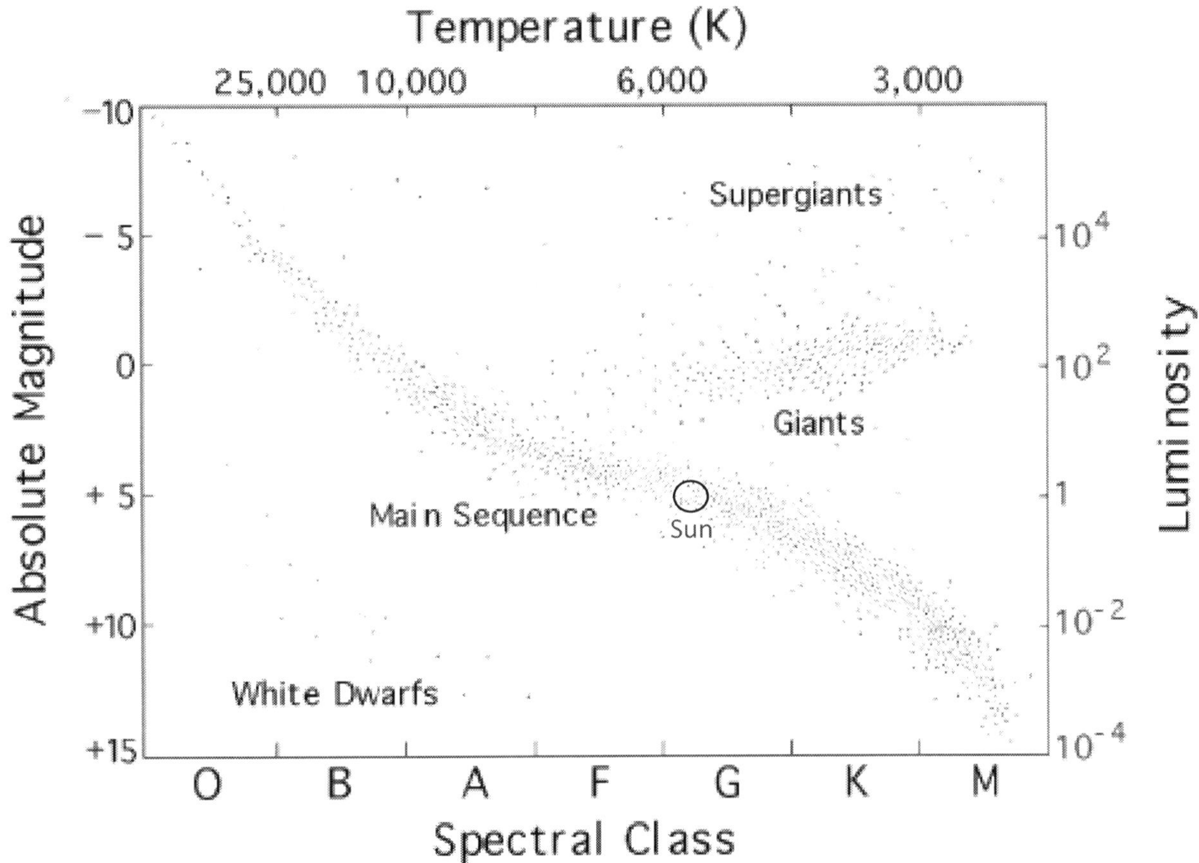

Directions:

Use the Hertzsprung-Russell diagram from NASA above to answer the questions below.

The **surface temperature** is on the top margin, the **class** of the star is on the bottom, absolute magnitude (how bright it would be if viewed from 10 parsecs or 32.6 lightyears away) is on the left margin, and **luminosity** (brightness) is in the right margin.

The most important factor in distinguishing one star from another – temperature, luminosity, size, and life cycle – seems to be the **star's mass**.

1) What classification is our sun?

2) What is our sun's surface temperature?

3) How does the size of the star determine its brightness?

4) How does the size of the star determine its surface temperature?

5) According to the diagram, what seems to determine the class of the star?

6) Does the class of the star determine its brightness? Explain.

7) The sun will one day become a red giant. Will the sun become brighter or darker? Explain how you know.

8) How will this affect life on Earth?

9) More massive stars have a much shorter life cycle (in millions of years) than our sun (10 billion years); in turn, less massive stars, like red dwarfs, have extremely long life cycles (possibly over 100 billion years). Should we expect to find well-evolved life (like on Earth) on a planet orbiting a giant or supergiant star?

10) If we could find it, where should we look for intelligent life outside of Earth (giving it more time to evolve)? Explain why.

Classifying Galaxies

Directions:

Use the **internet** to go to a NASA webpage using the URL: www.tinyurl.com/2p9xvay6. Use this page to answer the questions below.

1) What are the three main types of galaxies? Draw a picture of each.
    a.
    b.
    c.
2) What are the sub-classifications of the spiral galaxies with bulges? Draw a picture of each.
    a.
    b.
    c.
3) How are spiral galaxies with bulges different from ones with bars?

4) Draw pictures of some examples below.
   a. Bulges
   b. Bars

5) What are the sub-classifications of spiral galaxies with bars?
   a.
   b.
   c.

6) Why is there a number range for sub-classifications of elliptical galaxies? Draw pictures of them below.
   a.
   b.
   c.
   d.

7) What is the reason for the last classification of galaxies?

A Guide to the Milky Way Galaxy

Directions:

Use the **internet** to go to a NASA webpage at: https://www.nasa.gov/jpl/charting-the-milky-way-from-the-inside-out to answer the questions below. Use a **compass** to help you draw a map on the next page.

1) How would you describe where our solar system is located in the Milky Way galaxy?

2) Which arm are we on?

3) What is our spur?

4) Which arms of the galaxy are outside us?

5) Which arms are inside us?

6) Like the earth orbits around the sun in our solar system, our sun orbits a supermassive black hole at the center of our galaxy. The energy created from the friction of its distortion of the space-time continuum appears as a glow (we cannot see it because lots of gas is in the way blocking it) on the map. How many light-years (ly) are we from the center of the supermassive black hole?

7) How wide is the Milky Way?

8) How many stars are found in the Milky Way?

9) How did the scientist come up with this model of mapping our galaxy?

10) Is this a perfect map? What stops us from seeing our own galaxy?

11) Use the two maps from the NASA webpage to help you draw a map of the Milky Way galaxy on the rest of this page. Make sure to show and label all the important objects we know of. Use a compass to help you draw the circles showing the distance in lightyears from our sun.

Ten Things NASA Wants you to Know about the Universe

Directions:

Use the **internet** to go to the URL address: www.solarsystem.nasa.gov/solar-system/beyond/ to help you answer the questions that follow.

1) What was the big discovery Hubble made?

2) What kind of matter and energy makes up the universe? Give percentages of each.

3) What is the universe mostly made of?

4) How many galaxies are in our cosmic neighborhood?

5) Are there more planets or stars in the universe? Explain why.

6) What is the shape of the Milky Way?

    a. What other shaped galaxies are there?

7) Watch the video: Our Milky Way Galaxy: How Big is Space? How does light help us measure the size of the universe?

   a. How far is a light-second?

   b. How far is a light-minute?

   c. How far is a light-hour?

   d. How far is a light-Day?

   e. How far is a light-year?

8) Have we found other life out there?

   a. Do you think we will? Explain why.

9) What is at the center of the galaxy?

10) How many galaxies could be in our universe?

   a. How many were pictured in the tiny patch of sky on the web page?

The History of the Big Bang Theory

Directions:

Use the **internet** and your **textbook** to research the history of the Big Bang Theory of our universe. Then answer the following questions.

1) How does the Second Law of Thermodynamics show the universe had a beginning?

    a. Draw a graph of usable energy over time, showing the evidence.

2) What does Einstein's Theory of Relativity say about the relationship between matter and energy?

    a. Write his famous equation here.

3) Who used Einstein's equations to show mathematically the universe was expanding? (hint: there are two that are famous for this)

4) Who then saw physical evidence of this with a telescope?

    a. How did he use the Doppler Effect to show the universe is expanding?

5) Describe what the WMAP is and what it shows us.

6) What characteristic of the atom shows that the whole universe can fit inside a space smaller than an atom?

   a. What other characteristics of the atom show the energy of matter?

7) What are NASA and other scientists currently working on to help us learn more about the Big Bang?

8) What is quantum mechanics, and what does it show about the matter that makes up our universe?

9) What is the problem with Relativity and Quantum Mechanics?

10) Describe three other models that try to explain the origins of our universe.

a. 1st Model

b. 2nd Model

c. 3rd Model

11) Which of the models you researched today do you think best describes the origins of our universe? Explain why.

The Pixel of the Universe

Directions and Questions:

You will need a **golf ball**, a **bead**, and a **large field** or **parking lot**. **Looking at the materials and lab we will be using, what are the safety precautions we should take to protect ourselves and materials during this investigation?**

1) Walk out to a large field or parking lot, at least the size of a football field. Keep in mind the space you use still may be too small to be a scale model. You will make a model of a hydrogen atom with 1 proton and 1 electron.
2) On one edge, take a small red bead representing an electron and put it somewhere where you can see it (hang it on a fence or a tiny branch).
3) Walk at least 100 yards away; if you have more room, you can use that. Hold up the golf ball (a proton), read the information, and answer the following questions. Can you see the bead?

4) This distance is how far away the closest electron speeds around the proton. The speed approaches the speed of light. It moves so fast that it makes a ball the size of a football stadium. If you have ever seen a fan moving fast, does it look like a disk? But is it a disk?

   a. So we have to ask ourselves, the atoms, the pixels of our universe, look like solid balls (see the picture below), but are they balls? Explain why.

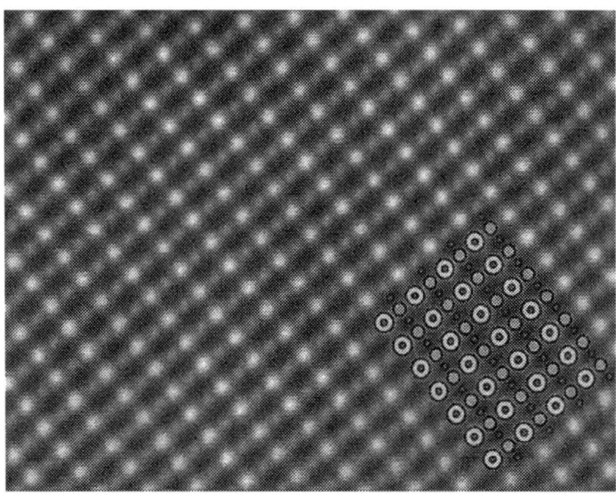

Atom resolution STEM image by Magnunor (Own work) CC BY-SA 4.0.

The illusion of the electron is a similar type of illusion. When an object moves close to the speed of light, time stops for that object. Quantum physics allows things without time to be everywhere they could typically be at the same time. So, why is the electron said to be everywhere in the electron cloud simultaneously?

5) What do you see between the proton and electron?

6) Quarks orbit inside protons, making them appear solid. Is there anything solid inside the atom? Explain.

7) If atoms are mostly empty space, why do you think they look solid?

8) What would happen to the electron if time were to stop?

9) If that would happen to the electron, what would happen to the atom?

10) What would happen to all atoms in the area where time stops?

11) This answer is why we call this the space-time continuum. You cannot have space without time. When time disappears, so does space. Quantum mechanics shows that objects outside of space and time create objects in space and time. The Theory of Relativity $E=mc^2$ says matter can be converted into energy, and energy can be converted into matter. The way the atom is structured shows us this is not only possible but also real. The whole universe has and can fit inside of what structure?

    a. We know all this is possible because the atom is mostly made up of what?

12) Why do we consider the atomic theory a theory? Why is it not a law or hypothesis?

WMAP

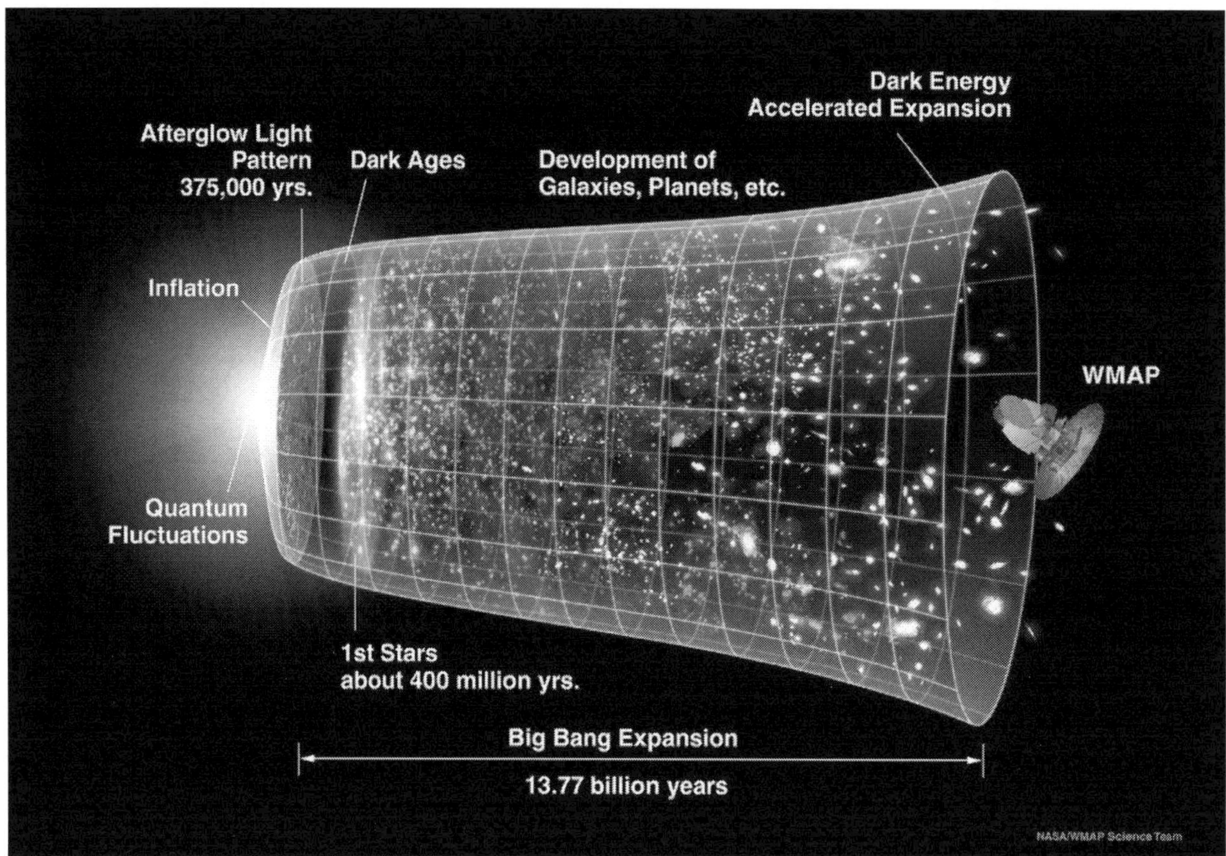

Directions:

Use NASA's image created by the Wilkinson Microwave Anisotropy Probe (WMAP) to answer the questions below.

1) How old is the universe?

2) What was the farthest this probe could look back(hint: Afterglow Light Pattern)?

3) How old was the universe when the first stars formed?

4) Describe how galaxies seem to change over time.

5) When does it look like the universe expansion started to accelerate?

6) What do physicists think is responsible for the accelerated expansion?

7) What is accelerated expansion of the universe?

8) What do you think this means for the future of our universe? Discuss with class.

9) Could there be a time in the future that other galaxies will not be visible to us from Earth (even with the most powerful telescopes)? Explain.

Virtual Investigations that go with the Development of the Universe

ExploreLearning.com:

Star Spectra

H-R Diagram

Orbital Motion – Kepler's Laws

Nuclear Reactions

Big Bang Theory – Hubble's Law

Solar System

Solar System Explorer

Comparing Earth and Venus

Phet.colorado.edu:

Blackbody Spectrum

Gas Properties

Gas Intro

Gravity and Orbits

Gravity Force Lab

Gravity Force Lab: Basics

Molecules and Light

Physicsclassroom.com:

Interactives:

Circular and Satellite Motion

Orbital Motion

The Value of g

Gravitation

Concept Builders:

Circular and Satellite Motion

Circular Logic

Case Studies – Circular Motion

Force and Free – Body Diagrams in Circular Motion

Universal Gravitation

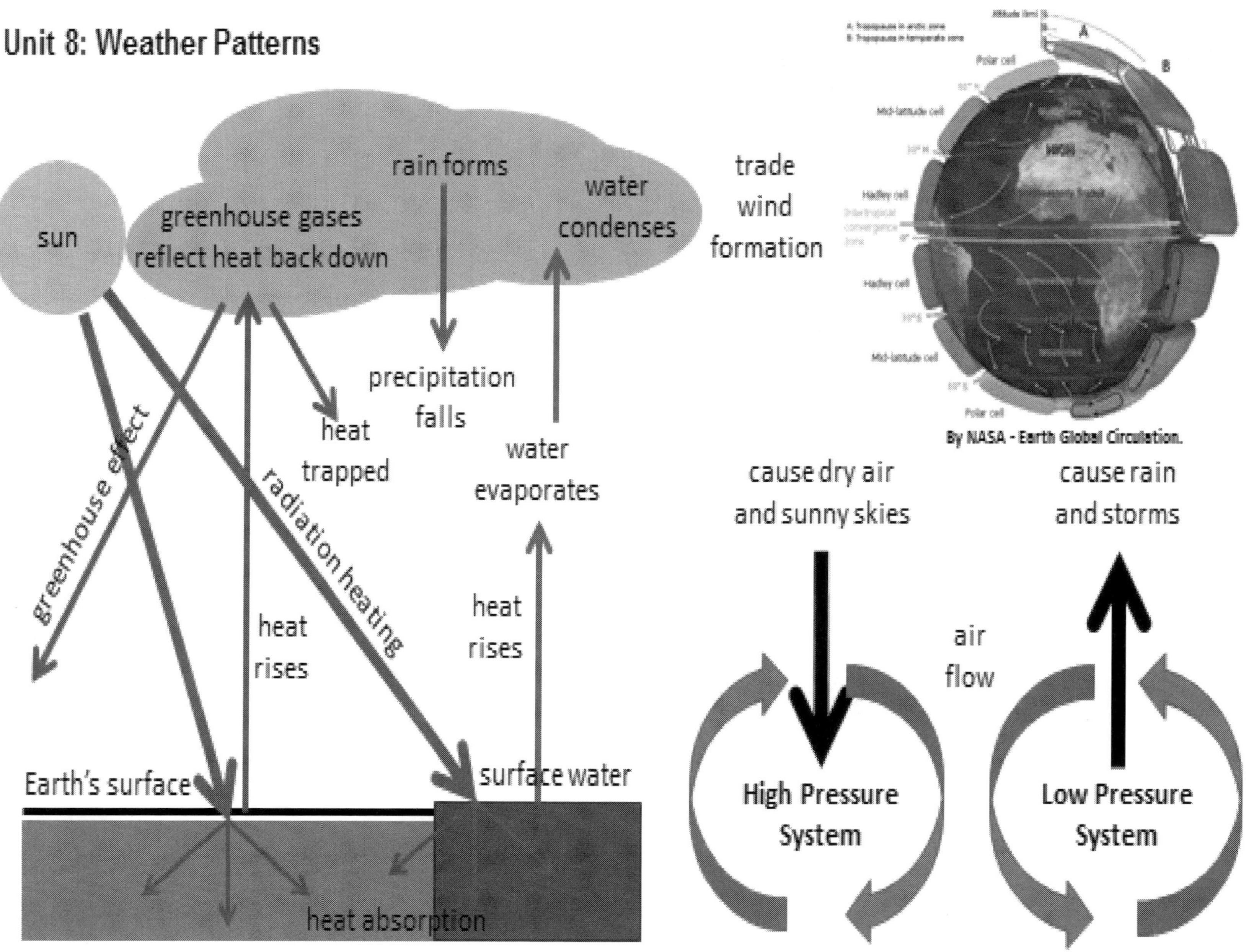
Unit 8: Weather Patterns
sun
greenhouse gases
reflect heat back down
rain forms
water
condenses
trade
wind
formation
precipitation
falls
heat
trapped
water
evaporates
greenhouse effect
radiation heating
heat
rises
heat
rises
Earth's surface
surface water
heat absorption
By NASA - Earth Global Circulation.
cause dry air
and sunny skies
cause rain
and storms
air
flow
High Pressure
System
Low Pressure
System

The Water Cycle

Directions:

Use the **internet** or your **textbook** to draw the water cycle below. Pay close attention to what directions the water moves to make it condense into clouds and rain.

Questions:

1) What direction does water have to move to condense into clouds and rain?

2) What happens to storms if forces make air move faster in that direction?

Seasons and the Tilt of the Earth

Directions:

You will need a **150-watt incandescent light source**, a **ring stand**, a **clamp** for the light source, a **globe**, **tape** that will hold the probe to your globe but not damage it, and a **temperature probe** attached to an **interface** connected to a **computer** with **Logger Pro**. **Looking at the materials and lab we will be using, what are the safety precautions we should take to protect ourselves and materials during the investigation?**

1) Set up your light on the ring stand to point at the middle of the globe. Place the bulb about 20 cm away from the globe.
2) Find your city on the globe and tape the temperature probe to the globe so that the probe's tip is over your city. Place the tape about 1 cm from the tip of the probe. Fold a piece of paper and wedge it under the back end of the temperature probe to keep it in contact with the surface of the globe.
3) Position the globe so it is winter for the northern hemisphere. The North Pole will be tilting away from the light. Open the folder for Earth Science with Vernier and file #29 Seasons.
4) Click the "Collect" button, then immediately turn on the lamp. It will collect data for 5 minutes. When the data collection stops, turn off the lamp. Choose store latest run from the Experiment menu. (Do not touch the bulb; it will be very hot!)
5) Let the bulb, globe, and temperature probe cool to the initial temperature in your first data run.
6) Position the globe so it is summer for the northern hemisphere. The North Pole will be tilting towards the light. Click the "Collect" button and then turn on the lamp. It will collect data for 5 minutes. When the data collection stops, turn off the lamp.
7) Click the Statistics button, and then click OK to display statistics boxes for both runs. Record the minimum and maximum temperatures for the winter and summer runs and calculate temperature change for both in Data Table 1.

Data Table 1

	Winter	Summer
Maximum Temperature (°C)		
Minimum Temperature (°C)		
Temperature Change (°C)		

Questions:

1) During winter in the northern hemisphere, is the light from the lamp able to hit the North Pole?

    a. How does this show there are 24 hours of the night during the winter?

2) Look at the South Pole. If the globe turns, will it ever go out of the light?

    a. How does this show there are 24 hours of the day during the summer?

3) How does this change when the Earth is positioned during the summer for the northern hemisphere?

4) Which season had a higher maximum temperature?

5) How does the temperature change for summer compared to winter?

6) Which season is the sunlight more direct?

7) What would happen to the temperature changes if the Earth were tilted more than 23.5°?

8) What causes the temperature and seasonal changes on Earth?

9) What other factors do you think could affect the weather in a region?

How Hurricanes Form

Directions:

Use the **internet** to go to a NASA URL address at https://tinyurl.com/4erbxnet. Use this web page to research how hurricanes form and answer the questions below.

1) Where do tropical cyclones form on Earth?

   a. Why do they form there?

2) What do hurricanes need to form?

3) Describe how a hurricane forms.

4) How does the eye form?

5) Why do hurricanes weaken when they go over land?

6) What determines the category of the hurricane?

7) Describe each category of hurricane listed here.

   a. 1

   b. 2

   c. 3

   d. 4

   e. 5

8) What causes the damage we see on land?

9) What allows us to better forecast where hurricanes will hit?

10) How do forecasters show where they predict the hurricanes will move to?

11) What may be causing the increase in the number and intensity of hurricanes?

12) What evidence did you see from NASA that this is happening?

13) What else is this causing?

14) What evidence did you see from NASA that what happens on one continent spreads to another?

15) Why is it important that all countries cooperate with each other regarding climate change?

Seeing Patterns in the Layers of the Atmosphere

Directions:

Table 1 contains the average temperature readings at various altitudes in the Earth's atmosphere. Plot this data on Graph 1. Connect adjacent points with a smooth curve. Be careful to plot the negative temperature numbers correctly. This profile provides a general picture of temperature at any given time and place; however, the actual temperature may deviate from the average values, particularly in the lower atmosphere.

Data Table 1

Altitude (km)	Temp (°C)	Altitude (km)	Temp(°C)
0	15	52	-2
5	-18	55	-7
10	-49	60	-17
15	-56	65	-33
20	-56	70	-54
25	-51	75	-65
30	-46	80	-79
35	-37	85	-86
40	-22	90	-86
45	-8	95	-81
48	2	100	-72

1) Label the different layers of the atmosphere and the separating boundaries between each layer (words are listed below).

2) Mark the general location of the ozone layer.

*Read the background information on page 178 and use it to place the eight words below on your graph in the correct locations:

troposphere, tropopause, stratosphere, stratopause, mesosphere, mesopause, thermosphere, and the **ozone layer**.

Graph 1

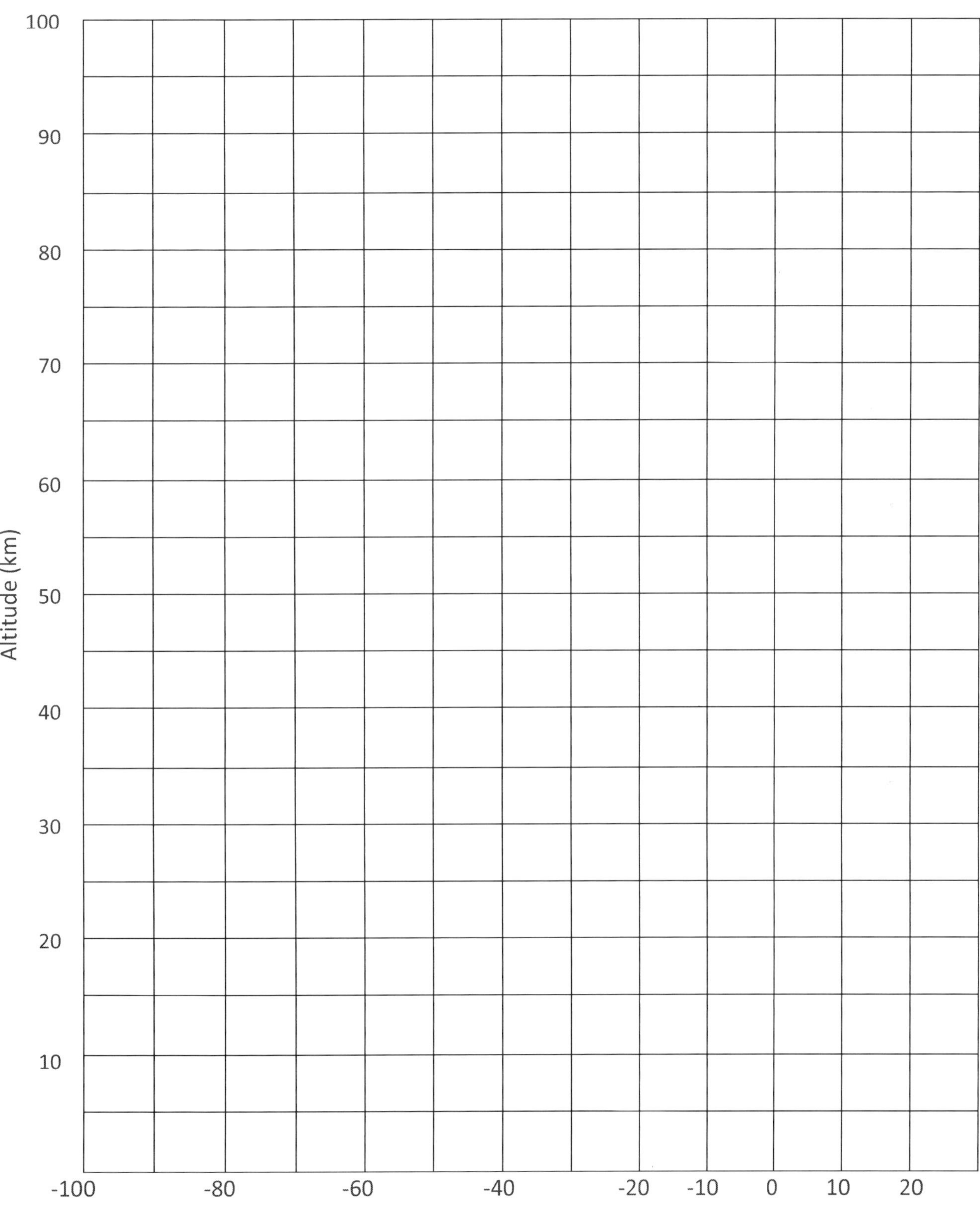
100
90
80
70
60
50
40
30
20
10
Altitude (km)
-100
-80
-60
-40
-20
-10
0
10
20
Temperature (°C)

BACKGROUND:

The atmosphere can be divided into four layers based on temperature variations. The layer closest to the Earth's surface is called the **troposphere.** The **stratosphere** is above this layer, followed by the **mesosphere**, then the **thermosphere**. The upper boundaries between these layers are the **tropopause**, the **stratopause**, and the **mesopause**.

Temperature variations in the four layers are due to how solar energy is absorbed as it moves downward through the atmosphere. The Earth's surface is the primary absorber of solar energy. Some of this energy is reradiated by the Earth as heat, warming the overlying **troposphere**. The **troposphere** is the lowest part of the atmosphere, where we live. It contains most of our weather, like clouds, rain, and snow. In this part of the atmosphere, the temperature gets colder as the distance above the Earth increases. The global average temperature in the **troposphere** rapidly decreases with altitude until the **tropopause**, the boundary between the **troposphere** and the **stratosphere**. It also contains about 75% of the air in the atmosphere and almost all water vapor, forming clouds and rain.

The temperature begins to increase with altitude in the **stratosphere**. This warming is caused by a form of oxygen called **ozone** (O_3) absorbing ultraviolet radiation from the sun. **Ozone** protects us from most of the sun's ultraviolet radiation, which can cause cancer, genetic mutations, and sunburn. Scientists are concerned that human activity contributes to a decrease in **stratospheric ozone**. Nitric oxide, which is in the exhaust of high-flying jets, and chlorofluorocarbons (CFCs), used as refrigerants, may contribute to **ozone** depletion.

At the **stratopause**, the temperature stops increasing with altitude. The overlying **mesosphere** does not absorb solar radiation, so the temperature decreases with altitude. Most of the meteors and rock fragments burn up in this layer.

At the **mesopause**, the temperature begins to increase with altitude, and this trend continues in the **thermosphere**. Here solar radiation first hits the Earth's atmosphere and heats it. Because the atmosphere is so thin, a thermometer cannot measure the temperature accurately, and special instruments are needed. This layer is relatively thin and is where the space shuttles orbited and the space station orbits today. A small change in energy can cause a large change in temperature in this layer. The temperature in this layer can rise to 1,500 °C or higher.

QUESTIONS:

1) What is the basis for dividing the atmosphere into four layers?

2) Does the temperature increase or decrease with altitude in the:

troposphere? _____________ stratosphere? ________________

mesosphere? _____________ thermosphere? _______________

3) What is the approximate height and temperature of the:

tropopause: _______________ _____________

stratopause: _______________ _____________

mesopause: _______________ _____________

4) What causes the temperature to increase with height through the stratosphere and decrease with height through the mesosphere?

5) What causes the temperature to decrease with height in the troposphere?

6) Describe the key characteristics of each atmospheric layer

Troposphere

Stratosphere

Mesosphere

Thermosphere

Global Wind Movement

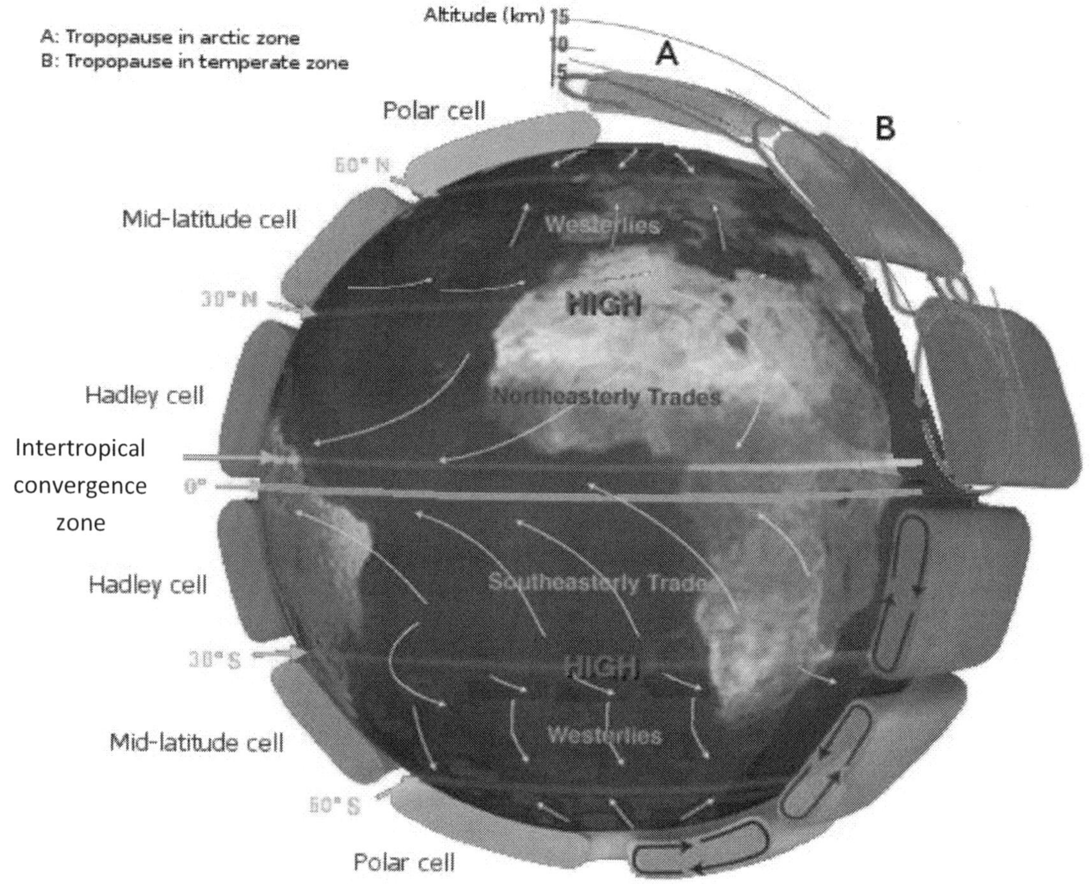

Picture by NASA - Earth Global Circulation.

Directions:

Use the NASA Picture above to help you answer the questions below on global wind patterns.

1) How many trade wind zones does the Earth have?

2) Which zone are you in on the Earth?

3) Which direction does your weather generally come from?

4) Where do the different trade winds move toward each other?

5) Where do the different trade winds move away from each other?

6) Look at the right side of the diagram. How do the convection cells change from the surface to the upper atmosphere?

7) Where is the hottest part of Earth?

8) When the sun heats the surface of this area which way does the air move (which direction does all hot air move)?

9) Heat likes to move from hot to cold areas. Where does the heat want to naturally move from the equator?

10) Using what you know about inertia, what does the atmosphere do when the Earth moves (hint: an object at rest wants to stay at rest)?

11) Which direction would this make the winds move across the Earth's surface?

12) The combination of the movement of the Earth and the direction heat wants to move gives us the trade winds; this is called the **Coriolis effect**. The Coriolis effect explains why high-pressure systems move in opposite directions in the northern hemisphere from the southern hemisphere. Why do cyclones (hurricanes) move in opposite directions in the northern and southern hemispheres?

Model Showing the Rotation of the Earth Stirring up the Atmosphere

Directions and Questions:

You will need a **spoon**, a **glass bowl** 2/3 filled with **water**, and some **pepper**. The pepper lets us better see the movement of the water when it is sprinkled in. **Looking at the materials and lab we will be using, what are the safety precautions we should take to protect ourselves and materials during the investigation?**

1) While the water is still, how does the pepper show it is moving?

2) In this model, the spoon will represent the movement of the Earth spinning on its axis, and the water will represent the atmosphere. Use the spoon to stir the water. How does the water move compared to the spoon?

3) Which is moving faster?

4) How does the pepper help you see the water moving?

5) Explain how Newton's 1st Law of Motion is involved with how the trade winds blow over the planet's surface because of the Earth's rotation.

6) What effect of the tradewinds is not included in this model?

A Local Weather Study

Directions:

You will need a **temperature probe**, a **humidity probe**, and a **UVB Sensor** attached to an **interface** connected to a **computer** with **Logger Pro**. **Looking at the materials and lab we will be using, what are the safety precautions we should take to protect ourselves and materials during the investigation?**

1) Go to a website that gives your local temperature and humidity. Each student should try a different site. We will test how accurate these sites are. Write this information in Data Table 1.
2) Take your computer, interface, and probes outside to test what the temperature and humidity actually are. Write this information in Data Table 1.
3) Subtract the difference between what the website said and what you actually measured. Write this information on the right side of Data Table 1.
4) Take the UVB sensor and point it at the sun. If possible, take measurements when there are no clouds in front of the sun. Then measure when clouds move in front of the sun. Look at the difference in the number of UV rays that come through the atmosphere with and without clouds. Write this in Data Table 1, continued on the next page.
5) Compare your results with the rest of the class to see which websites were most accurate and answer the following questions.

Data Table 1

The website used: ____________________	Website Measurements	Actual Measurements	Difference
Temperature			
Humidity			

UVB Reading	No Clouds:	With Clouds:	Difference:

Questions:

1) Describe the location where you recorded your measurements. Include observations such as:
    a. Is the spot open? Are there buildings, trees, or other objects that could have affected your measurements?

    b. What is the ground cover?

    c. Are there any living organisms in the immediate area?

2) Which website did the class find to be the most accurate?

3) Why do you think that website's measurements were so close to yours?

4) What do you think will happen to the humidity measurements at different times of the day?

a. Why do you think this will happen?

5) What do you think will happen to the temperature measurements at different times of the day?

   a. Why do you think this will happen?

6) What part of the atmosphere blocks harmful UV rays from the sun?

7) Why is it important to know how many UV rays come through the atmosphere?

Relative Humidity

Directions:

You will need a cut piece of a **shoelace**, **tape**, a **beaker** of **water**, and two **temperature probes** attached to an **interface** connected to a **computer** with **Logger Pro**. **Looking at the materials and lab we will be using, what are the safety precautions we should take to protect ourselves and materials during the investigation?**

1) The first temperature probe will be the dry probe inserted into Channel 1 on the interface. Take the cut shoelace and tape it to the tip of the second temperature probe; insert this into Channel 2. This probe will be the wet probe.
2) Place the second temperature probe in the beaker of water to wet the shoelace.
3) Hold the first temperature probe in one hand, take the second temperature probe out of the water, and gently wave it back and forth to help the water evaporate. Wait for the temperature to even out as it drops on the second temperature probe (this could take a few minutes).
4) When the two probes have steadied their temperatures, write them in Data Table 1.
5) Repeat the procedure for #s 2-4 for two different locations of your teacher's choice.
6) Subtract the wet probe temperature from the dry probe to find the temperature differences at each place. Write this in Data Table 1.
7) Use the temperature difference with the dry probe temperature in Data Table 2 to find each humidity at the different places. Place these humidity values at the bottom of Data Table 1.

Data Table 1

	Classroom	Site 2:	Site 3:
Dry Probe Temperature (°C)			
Wet Probe Temperature (°C)			
Temperature Difference (°C)			
Relative Humidity (%)			

Data Table 2

	Dry Probe Minus Wet Probe Temperature(°C)									
Dry Probe Temperature	1	2	3	4	5	6	7	8	9	10
10°C	88	77	66	55	44	34	24	15	6	
11°C	89	78	67	56	46	36	27	18	9	
12°C	89	78	68	58	48	39	29	21	12	
13°C	89	79	69	59	50	41	32	22	15	7
14°C	90	79	70	60	51	42	34	26	18	10
15°C	90	80	71	61	53	44	36	27	20	13
16°C	90	81	71	63	54	46	38	30	23	15
17°C	90	81	72	64	55	47	40	32	25	18
18°C	91	82	73	65	57	49	41	34	27	20
19°C	91	82	74	65	58	50	43	36	29	22
20°C	91	83	74	67	59	53	46	39	32	26
21°C	91	83	75	67	60	53	46	39	32	26
22°C	92	83	76	68	61	54	47	40	34	28
23°C	92	84	76	69	62	55	48	42	36	30
24°C	92	84	77	69	62	56	49	43	37	31
25°C	92	84	77	70	63	57	50	44	39	33
26°C	92	85	78	71	64	58	51	46	40	34
27°C	92	85	78	71	65	58	51	46	40	34
28°C	93	85	78	72	65	59	53	48	42	37
29°C	93	86	79	72	66	60	54	49	43	38
30°C	93	86	79	73	67	61	55	50	44	39

Questions:

1) How did the wet probe temperature compare with the dry probe temperature at each site? Explain why.

2) Which site had the highest relative humidity?

3) Which site had the lowest relative humidity?

4) Explain what could cause the relative humidity to be different at these locations on the same day.

5) How do you think the humidity would change during the day at different times outside?

6) Compare the relative humidity values outside at sunny and shaded sites under trees.

7) What could be some sources of error in this investigation?

Temperature Inversions

Directions:

Use the **internet** to research temperature inversions, then answer the following questions.

1) What are temperature inversions?

2) What are some examples of temperature inversions, and where can we find them?

3) What are the short-term effects of temperature inversions?

4) What are some long-term effects of temperature inversions?

5) How do they affect El Nino and La Nina oscillations?

6) How do they affect the polar ice caps and glacial melting?

7) How do they change the ocean surface temperatures?

Virtual Investigations that go with Weather Patterns

ExploreLearning.com

Weather Maps – Metric Gizmo

Relative Humidity Gizmo

Coastal Winds and Clouds Gizmo

Weather Maps

Observing Weather (Metric) Gizmo

Observing Weather (Customary) Gizmo

Comparing Climates (Metric) Gizmo

Comparing Climates (Customary) Gizmo

Hurricane Motion Gizmo

Seasons Around the World

Seasons in 3D

Seasons: Why do we have them?

Summer and Winter

Water Cycle

Convection Cells

Greenhouse Effect

Unit 9: Influences on Climate Change

Sun
solar radiation
climate factors
ocean circulation patterns
atmospheric patterns
regional changes
Climate Change

Weather Patterns	
hurricanes	clouds
seasons	wind

air quality
atmospheric pollution
atmospheric changes
Human Activities
Natural Events

burn fossil fuels	pollution emissions	fires	volcano
population growth municipal development		meteor impacts	

Regional and Global Consequences

Ecological Consequences
acid rain ozone depletion smog increase greenhouse gases

Composition of the Atmosphere

1) Use the **internet** to list the top 5 chemicals and their % in the atmosphere (include how much the water range can be).

2) Describe the Earth's first atmosphere and tell how it may have changed to what it is today. **Hint:** There were 3 atmospheres. Tell what the first atmosphere was composed of, then what happened to cause the second, then what happened to cause the one we have today.

3) When nature spent billions of years locking the old atmospheric chemicals in the Earth's crust, what do you think would be the effects of digging them up and burning them, putting them back into the atmosphere?

The Greenhouse Effect

Directions:

You will need two **plastic tubs** (shoebox size) **painted black** on the inside, a **"Press'n Seal" sealing wrap**, two **temperature probes** attached to an **interface** connected to a **computer** with **Logger Pro**, and a **light source** like an incandescent lamp or some other lamp that gives off heat. **Looking at the materials and lab we will be using, what are the safety precautions we should take to protect ourselves and materials during the investigation?**

1) Make sure your tubs are both painted black on the inside, and a hole is poked through the end of each tub, big enough for a temperature probe to fit in snugly. Push the temperature probes through the holes of both tubs.
2) On the first tub, make sure the Press'n Seal is secured to the tub's opening and the temperature probe is plugged into channel 1 of the interface.
3) The other tub needs to remain open, and the temperature probe plugged into channel 2.
4) In Logger Pro, open the folder Earth Science with Vernier and file #24 Greenhouse Effect.
5) **Hypothesis:** which tub do you think will heat up faster?

6) Make sure your light source is equal distance from both tubs. Press "Collect" in the Logger Pro and turn on your lamp.
7) Monitor the time; when **5 minutes have passed, turn off the light**. Data will continue to be collected.
8) At the **10 minutes, turn the lamp back on**. Data collection will continue until 15 minutes. At 15 minutes, Data collection will stop.
9) Look at the data collected in the Logger Pro and fill in Data Table 1 for the initial temperature, the temperature at 5 minutes, the temperature at 10 minutes, and the temperature at 15 minutes.
10) Then subtract the temperatures between Probe 1 and Probe 2 to get the temperature differences at different times. Write this information on the right side of Data Table 1.

Data Table 1

	Probe 1 Greenhouse	Probe 2 Control	Temperature Difference
0 Minute Temperature (°C)			
5 Minute Temperature (°C)			
10 Minute Temperature (°C)			
15 Minute Temperature (°C)			

Questions:

1) When the lamp was on, which tub heated faster?

2) Give a possible explanation for your answer to number 1.

3) When the lamp was off, which tub cooled faster?

4) Give a possible explanation for your answer for number 3.

5) How does this information show how temperatures could increase over time because of the greenhouse effect?

6) What are we doing to the atmosphere to increase greenhouse gases' effects to cause global warming and climate change?

7) How does this information show temperature could decrease over time without the greenhouse effect?

8) What do you think happened to the atmosphere to allow ice ages to happen on the Earth?

9) When do you think the temperature will get the coldest at night: when it is a cloudy night or when there are no clouds at night? Explain why.

10) The habitability of the Earth is a result of a delicate balance of the greenhouse effect. How/why is this statement true?

11) Explain why a closed automobile heats up in the sun.

12) Why do you not leave your child or pet in the car on warm days when the car is parked and turned off?

13) What could be sources of error in this investigation?

14) Do you think the Greenhouse Effect is a hypothesis or a theory? Explain why.

Climate and Greenhouse Gases: Data Table

Directions:

Look at the Data Table below and build line graphs on the following pages showing the trends in temperature and carbon dioxide data. Then compare the trends in the graphs together, answering the questions that follow. This data was collected through ice core samples where the atmosphere was trapped inside the snow and then compressed into ice. When we melt the different layers of ice, we can accurately measure Carbon Dioxide levels in those layers. Also, by measuring the number of Oxygen isotopes, we can get an accurate temperature of that time.

Data Table 1

Years Before Present (x 1000)	Local Temperature Change (°C)	Carbon Dioxide (ppm)
160	-9	190
150	-10	205
140	-10	240
130	-3	280
120	1	278
110	-4	240
100	-8	225
90	-5	230
80	-6	220
70	-8	250
60	-9	190
50	-7	220
40	-8	180
30	-7	225
20	-9	200
10	-2	260
0 (Year about 1850)	-0.5	280
0 (Year about 2002)	-	371

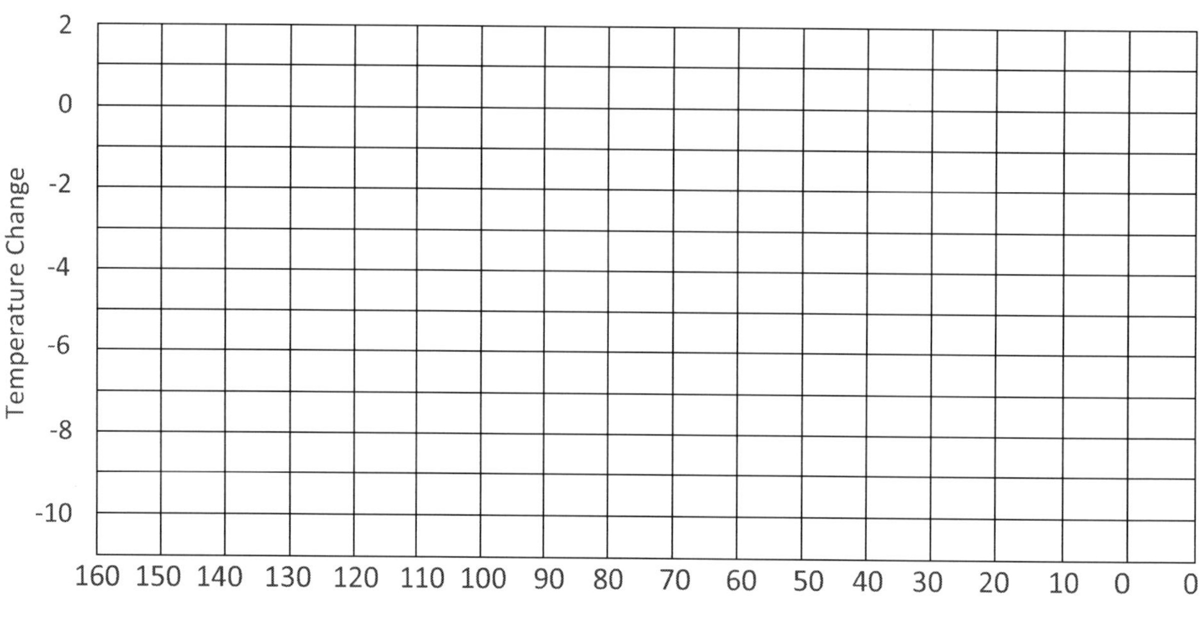

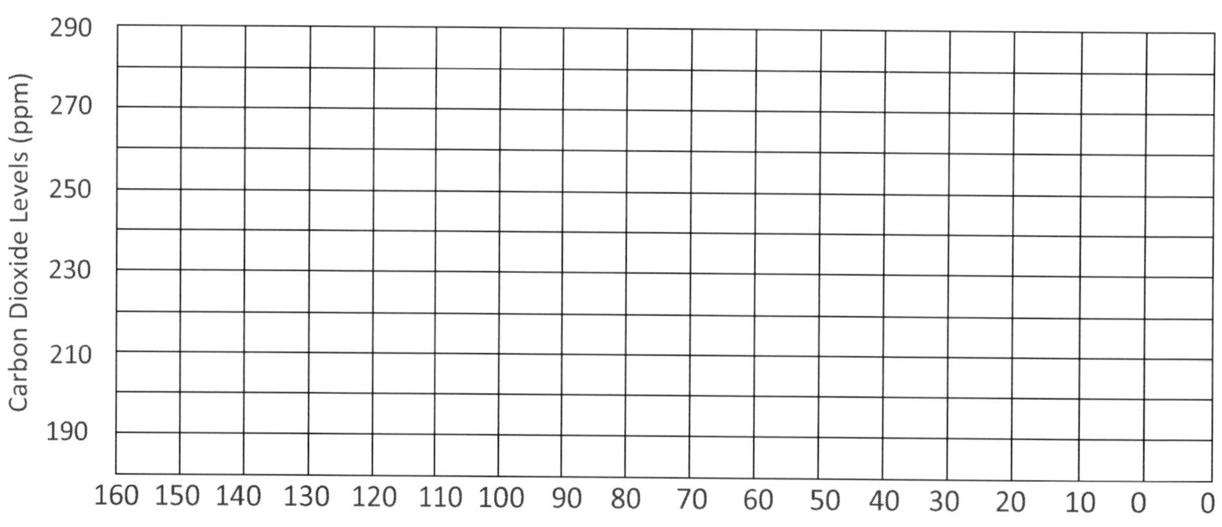

Questions:

1) Do you see any similarities between the temperature graph and the Carbon Dioxide levels graph?

2) Which graph literally went off the chart and where?

3) What could this mean for the other graph in the near future?

4) Does this evidence show Carbon Dioxide is causing the temperature change? Explain.

5) How might this affect our future climate and where people live if the Earth is getting warmer?

6) How could this affect the evolution of life?

7) Can anything be done to prevent increasing temperatures? If so, what are they?

8) Do you think Global warming is a hypothesis or a theory? Explain why?

Carbon Dioxide and Population

Directions:

Graph the information from Data Table 1 onto Graph 1, then answer the questions below. Use the left Y-axis to create a line graph for the Population and the right Y-axis to create a line graph for the carbon dioxide emissions. Make sure you make a key showing which line on the graph is the population and which line is the carbon dioxide emissions.

Data Table 1

Year	Population (in millions)	Carbon Dioxide Emissions (in metric tons)
1750	790	11
1800	980	29
1850	1260	198
1900	1650	1,982
1950	2520	5,982
2000	6060	25,620

Graph 1

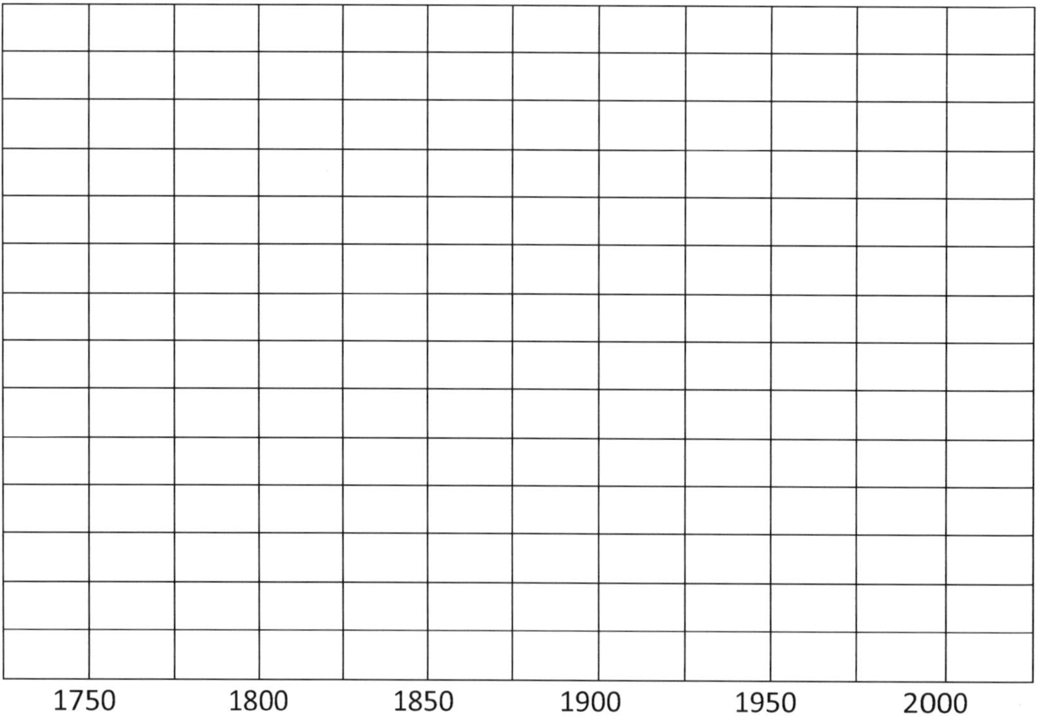

Questions:

1) What happens to the carbon dioxide emissions as the population of humans rises?

2) Why do you think this happens?

3) How has the source of carbon dioxide emissions changed over the years?

4) Is there any way we can lower those emissions? If so, how?

5) What kind of culture change do we need if we are going to lower the carbon dioxide levels?

6) What is happening to the Earth because of the rising carbon dioxide levels?

7) How does this impact humans now?

8) How could this impact humans in the future if we do not change the trends in the data?

9) How could this impact the evolution of life in the future if the trends continue?

Carbon Emissions

Directions:

1) Your teacher will assign you a family scenario. Use it as you follow the directions to fill in your Carbon Emissions in Data Table 1. Once done, check with another person with the same family as you to check for errors. Once done, follow the directions again to fill out a carbon emissions Data Table 1 for your family. Use a five days a week work week, 52 weeks a year, four weeks in a month, 12 months a year, and 365 days a year. Everyone will not use all parts of the calculation. Only use the ones that apply to your family.
2) To comply with the **Kyoto Protocol**, we should only produce 5.4 tons of CO_2 per person per year. To make a difference to help control global warming, we could produce no more than 2.35 tons of CO_2 per person per year.
3) Number of miles traveled by car #1 each year_______, divided by average miles per gallon = _______ gallons of gas multiplied by 22 pounds of CO_2/gallon of gas = _______ pounds of CO_2 from Car 1. Write this number for Car #1 in Data Table 1.
   a. Repeat #3 for additional cars and any other fuel motor vehicles, including motorcycles, boats, etc.
4) Number of miles of air travel per year (all household members) _______, multiplied by 0.9 pounds of CO_2/mile of air travel = _______ pounds of CO_2. Write this in Data Table 1.
5) Number of miles traveled on mass transit (bus, train) _______, multiplied by 0.5 pounds of CO_2/mile of mass transit travel = _______ pounds CO_2. Write this in Data Table 1.
6) The number of miles traveled by taxi, limo, UBER _______, multiplied by 1.5 pounds of CO_2/mile in a taxi, limo, or UBER = _______ ponds of CO_2. Write this in Data Table 1.
7) Kilowatt-hours of electricity used per year _______, multiplied by 1.5 pounds of CO_2/kilowatt-hour = _______ pounds of CO_2. Write this in Data Table 1.
8) Therms of natural gas per year _______, multiplied by 11 pounds CO_2/therm = _______ pounds of CO_2. Write this in Data Table 1.
9) Gallons of propane or bottled gas per year _______, multiplied by 13 pounds CO_2/gallon = _______ pounds of CO_2. Write this in Data Table 1.
10) Add up the pounds of CO_2 emitted by the household. Write this in Data Table 1.
11) Divide the total by the number of people in the household = _______ pounds of CO_2 emitted per person in one year. Divide the total by 2000 pounds to determine the number of tons of CO_2 per year per person. Write this in Data Table 1.
12) Find the tons of CO_2 per year per person for the other three families from the other groups in the class.
13) Repeat the procedure for #2-11 for your own family. Write this into Data Table 3.

Family #1

Ralph and Brenda are both in their early 30s and live inside the Loop in a 1200 square-foot condo. They both work downtown and take the bus to work. She commutes about 8 miles per workday on the bus, and he travels about 7 miles per day on the bus. Since they do not need cars for a commute, they own one car, an Acura TL, that gets 23 miles per gallon. They drive the car on average 7 miles per day because they live close to dining and recreation. They fly an average of 3000 miles a year each to visit family and vacation for the year. Their condo uses an average of 950 kilowatts of electricity each month, and all appliances, including heating and air conditioning, are electric.

Family #2

Bill and Ellen are in their early 40's with one child, Mark. They live in the Heights in an 1800 square-foot home that they remodeled to be energy efficient. They are close to everything they need, including a grocery store and a farmer's market. Both work; however, they have an office inside the house, allowing Bill to work from home three days a week. The other two days a week, he commutes to his nearby office. Ellen is an artist who works out of a loft in their house. They own energy-efficient cars. He drives a Honda Civic Hybrid that gets about 40 miles to the gallon, and she drives a Toyota Prius that gets 45 miles to the gallon. On average, Bill drives 30 miles each week, and Ellen drives 40 miles each week. Mark attends a nearby school and rides his bike to school. Despite their house being fully electric, they only use an average of 350 kilowatts of electricity per month because they have solar panels on the roof of their house that generate part of the electricity for their household.

Family 3

Matt and Debbie, who are in their 40's, live in the suburbs with their three children. Their house is 2000 square feet and uses all electricity. On Average, they use 1200 kilowatts of electricity per month. Debbie works part-time and drives a minivan that gets 20 miles per gallon. Between work, running family errands, and driving children, she drives about 30 miles per day. Matt has to drive more for his work and sometimes drives to other work sites, so he puts about 60 miles per day on his pick-up truck that gets 14 miles to the gallon. To relax, Matt enjoys grilling on their outdoor propane grill. On average, he uses 1.5 gallons of propane each month.

Family 4

Ben and Dawn live in a 4,000 square-foot house with all the amenities, including a heated pool, a family room with a built-in home theater system, and several televisions throughout the house, including one upstairs solely for the kids' Wii. With a house this large and with so many electrical devices, they use on average 5000 kilowatts of electricity per month. The house has a natural gas stove and oven, outdoor grill, water heater, and furnace. These appliances use on average 65 therms of natural gas each month. Dawn drives a full-sized SUV that gets 18 miles to the gallon, which she uses to take the kids to their private school, run errands for herself and her family, and take kids to their after-school activities. On average, Dawn drives about 60 miles per day. Ben enjoys driving his Hummer H3, which gets about 13 miles to the gallon. He has a fairly long commute to work, so he drives about 50 miles per day. Twice a year, they go on vacation as a family. Each family member travels by airplane an average of 7000 miles each year for these vacations. Ben has to travel for business once a month. He takes a taxi to and from the airport, which is 25 miles from their house. He flies to Boston, which is 1600 miles away.

Data Table 1: Family #_____

Source per Year	Car 1	Car 2	Air Travel	Mass Transit	Taxi UBER	KWh	Therms Natural Gas	Propane	Total	Total per Person	Tons per person
Pounds CO2											

Data Table 2: Family Comparison

Family	Family 1	Family 2	Family 3	Family 4
Tons of CO2 per Person				

Data Table 3: My Family

Source per Year	Car 1	Car 2	Air Travel	Mass Transit	Taxi UBER	KWh	Therms Natural Gas	Propane	Total	Total per Person	Tons per person
Pounds CO2											

Questions:

1) What was the largest contributor to the family you were assigned carbon emissions?

2) Would this family's habits help or hinder efforts to control global warming?

3) Compare this family to the other families. What are the biggest differences?

4) Compare your family in Data Table 3 to families 1-4. Which family does your family closely resemble?

5) Is your family Kyoto compliant?

6) Would your family help or hinder efforts to control global warming?

7) What are some things your family could do to help efforts to control global warming?

Climate Change

Directions:

Use the **internet** to research climate change and answer the questions below.

1) What is climate change?

2) What are the causes?

3) What is the impact of climate change on polar ice caps and glaciers?

4) How would this affect ocean currents?

5) How does it affect the surface temperatures of the Earth?

6) How is it impacting humans worldwide, and what can we expect in the future if this is happening?

7) What evidence do we have that it is happening?

8) What are the arguments against climate change?

9) Who are the people (occupations) studying climate change?

10) Do you think it is man-made or natural? Explain in detail why with evidence for your opinion.

How is Life Allowed on Earth?

Directions:

Use your **textbook** and **internet** as resources and what you have learned to explain the characteristics of the Earth that allow life to exist on it. Use this to answer the questions below:

1) What kind of sun/star do we need?

2) How far away does a planet need to be?

3) What effect does the rotation of a planet have on the ability for life?

4) How would a tilt of the axis affect the planet?

5) How does the moon help the Earth support life?

6) How does the atmosphere support life?

7) How does the magnetic field help the atmosphere on Earth?

8) What materials are needed for life to exist and thrive?

9) How would photosynthesis and aerobic respiration use these resources to help life thrive?

10) What can we learn about Venus and Mars to give us clues about the Earth?

11) What is the Goldilocks Zone, and how does that fit in?

12) What things in the universe could destroy all life on Earth?

13) What protections are we getting from our solar system?

14) How complicated is the structure of life?

15) What does it take to make one cell the basic unit of life?

16) After researching this, how fragile is life on Earth, and how likely do you think we will find life on other planets?

   a. Would it look like life on Earth? Explain why.

Natural and Manmade Disasters

Directions:

Use the **internet** to fill in the following charts while researching natural and manmade disasters.

Natural Disasters

Name	What it does	Disasters that Cause it	Disasters it can Cause	Environmental Impacts	Impacts on Humans	Short Term Effects	Long Term Effects
Hurricane							
Flood							
Tornado							
Earthquake							
Volcanic Eruption							
Tsunami							
Mudslide/ Avalanch							
Drought							
Wildfire							
Deforestation							

Manmade Disasters

Name	What it does	Disasters that Cause it	Disasters it can Cause	Environmental Impacts	Impacts on Humans	Short Term Effects	Long Term Effects
Oil Spill							
War							
Plane Crash							
Nuclear Reactor Meltdown							
Nuclear Bomb							
Mining Accident							
Municipal Development							
Fires							
Global Warming							
Fracking							

Digital Presentations of Worldwide Disasters

Directions:

Use the **internet** and your **textbook** to create two presentations in a digital format designated by your teacher (examples: Animoto, Prezi, PowerPoint, or Word). Make one for a Natural disaster and one for a Manmade Disaster.

1) Follow the rubric below and include pictures, videos, information from your charts, and any other information you think is important or interesting.

Poster Title: ______________________________

Rubric

Scores range from 0-10 in each box.

Title: Include the name of the disaster and the author's name 0, 5, or 10	Accuracy of the information 0-10	What happens in the disaster 0-10	What other disasters can cause it 0-10
What other disasters it can cause 0-10	Environmental Impacts of the disaster 0-10	Impact of disaster on humans 0-10	Short term effects 0-10
Long term effects 0-10	The overall feeling of the presentation and its appearance 0-10	**Total Points** 100 points possible	

Student's Name: ____

Nonrenewable Resources Chart

Directions:

Use the **internet** and your **textbook** to research and fill out this chart on nonrenewable energy resources.

Type	How do we obtain and transport it	How we use it	How it affects the environment
Petroleum			
Coal			
Natural Gas			
Nuclear			

Questions:

1) What are the advantages of using petroleum?

2) What are the disadvantages of using petroleum?

3) What type of reaction goes into burning petroleum products for energy?

    a. Are there any products from this reaction that are harmful to the environment? If so, how are they harmful?

4) What are the advantages of using coal?

5) What are the disadvantages of using coal?

6) What type of reaction goes into burning coal for energy?

    a. Are there any products from this reaction that are harmful to the environment? If so, how are they harmful?

7) What are the advantages of using natural gas?

8) What are the disadvantages of using natural gas?

9) What type of reaction goes into burning natural gas?

    a. Are there any of the products from this reaction that are harmful to the environment? If so, how are they harmful?

10) What are the advantages of using nuclear energy?

11) What are the disadvantages of using nuclear energy?

12) Are there any products from nuclear reactions that are harmful to the environment? If so, how are they harmful?

13) Describe some careers involved with the exploration, extraction, production, and disposal of these resources.

Renewable Resources Chart

Directions:

Use the **internet** and your **textbook** to research and fill out this chart on renewable energy resources.

Type	How we obtain and transport it	How we use it	How it affects the environment
Wind			
Solar			
Hydroelectric			
Geothermal			

Questions:

1) What are the advantages of using wind energy?

2) What are the disadvantages of using wind energy?

3) What are the advantages of using solar energy?

4) What are the disadvantages of using solar energy?

5) What are the advantages of using hydroelectric energy?

6) What are the disadvantages of using hydroelectric energy?

7) What are the advantages of using geothermal energy?

8) What are the disadvantages of using geothermal?

9) Describe some careers involved with the production of energy using these resources.

Virtual Investigations that go with Influences of Climate Change

ExploreLearning.com

Greenhouse Effect

Coral Reefs 1 – Abiotic Factors

Coral Reefs 2 – Biotic Factors

Ocean Carbon Equilibrium STEM Case

Ocean Carbon Equilibrium Handbook

Seasons Around the World

Observing Weather

Weather Maps

Convection Cells

Weathering

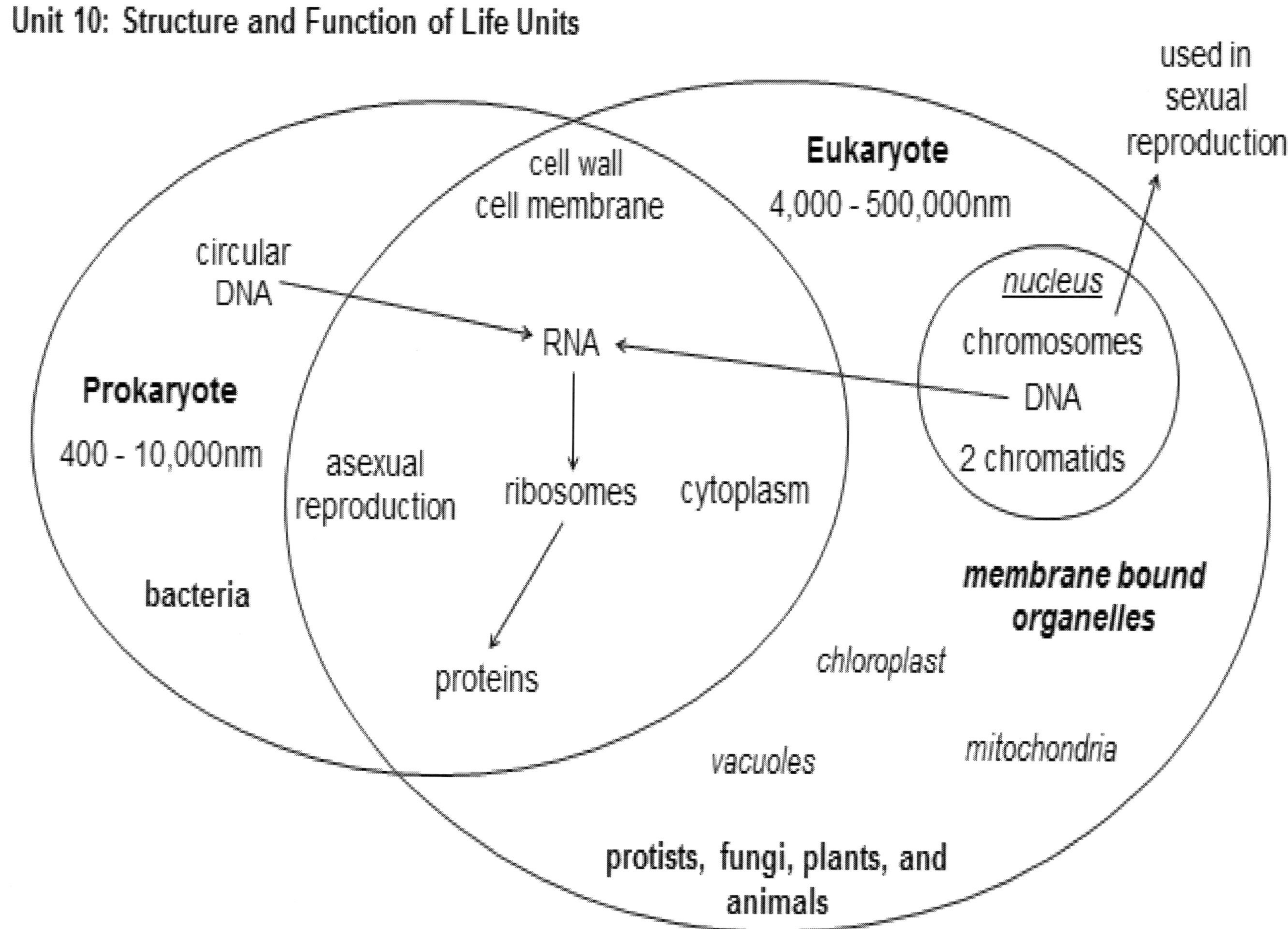
Unit 10: Structure and Function of Life Units
cell wall
cell membrane
Eukaryote
4,000 - 500,000nm
used in
sexual
reproduction
circular
DNA
RNA
nucleus
chromosomes
DNA
2 chromatids
Prokaryote
400 - 10,000nm
asexual
reproduction
ribosomes
cytoplasm
bacteria
membrane bound
organelles
chloroplast
proteins
vacuoles
mitochondria
protists, fungi, plants, and
animals

Membrane Models

Directions and Questions:

You will need a **large beaker of water**, **food coloring**, **dishwashing liquid** or some **soapy bubble mixture**, **long twisty ties**, a **small shallow dish**, **toothpicks**, a **strainer** or **colander**, a **small beaker of water**, **salt**, **marbles**, **dry beans**, a **small tub** or **bucket**, and two **pieces of paper**. **Looking at the materials and lab we will be using, what are the safety precautions we should take to protect ourselves and materials during the investigation?**

Part 1 Diffusion and Osmosis

1) Take the large beaker of water and drop a couple of drops of food coloring in the water. How do you see the food coloring move?

2) When did it seem to stop moving?

3) You just watched the process of **diffusion**. And the diffusion of water is **osmosis**. These are examples of **passive transport**. The next two sections of the lab will show you how some materials can passively move across a cell membrane and others cannot.

Part 2 Semipermeable by Material

4) Cover the bottom of the small dish with soap. Take two twisty ties. Make a large loop with the first tie; with the second tie, make a much smaller loop. Bend the rest of both ties up at a 90° angle to use as a handle to dip into and pull out of the soap.
5) Now dip the loops into the soap and pull it out. Notice it produces a thin soapy membrane that stretches across the loop. Which loop holds the membrane on it longer? This model should show us why cells cannot grow bigger; the membrane cannot hold together a large cell.

6) Now dip the small loop into the soap and pull it out with a soapy membrane. Take a toothpick and poke it through the membrane. What happens to the membrane?

7) This time repeat the procedure in #6 but this time, wet the toothpick with water, then gently poke it through the soapy membrane. What happened to the membrane this time?

8) This time repeat the procedure in #7 but take a new toothpick, cover one end of it with soap, and then gently poke it through the soapy membrane. What happened to the membrane?

9) Make sure the small loop is lined with soap. Dip the large loop into the soap and pull it out with a soapy membrane. Now try to pass the small loop through the larger loop with the membrane while holding it parallel to the ground. What do you notice? This process makes an opening act as a protein channel, allowing materials to move in and out of the cell membrane.

10) What was used as a model of a cell membrane in this part of the investigation?

Part 3 Semipermeable Membrane by Size

11) Observe the colander; what does it have all over it that allows some things to pass through and other things not pass through? These represent the protein channels in the cell membrane.

12) Take the water, salt, marbles, and dry beans and see which substances can pass through as you pour them in over a small tub or bucket. Which materials were allowed to pass through?

13) Which materials were not allowed to pass through?

14) Why did those materials not pass through? This model shows why many molecules do not pass through in and out of cell membranes; this is why cell membranes are semipermeable, allowing some things to pass through and stopping others.

15) What was used as a model of the cell membrane in this part of the investigation?

Part 4 Active Transport

16) When molecules are too big to pass through, many substances can be recognized by the cell membrane and act as your classroom door. A flat piece of paper can slide under the door. Slide the flat paper under the door. Did it go under the door?

17) A wadded paper cannot pass under the door. Wad the second paper and try to have it pass under the door. Was the wadded paper able to pass under the door?

18) But a large protein (like me) can trigger the doorknob to open it up to allow me to pass through. There are structures outside the cell membrane that trigger **endocytosis** allowing large substances to enter the cell.
    a. When it eats, it is called **phagocytosis,** forming a food vacuole.
    b. Pinocytosis is when the cell drinks a large amount of water all at once, forming a water vacuole.

c. When I pass through the door, what type of endocytosis is this modeling?

19) How do you think a large protein made inside the cell is allowed to pass out of the cell during **exocytosis** (hint: remember the soap)?

20) Was energy needed to open the door?

21) What was used as a model of a cell membrane in this part of the investigation?

22) You can think of **passive transport** as moving with an escalator. Do you have to use any energy to move with the escalator?

23) You can think of **active transport** as trying to move against an escalator. Do you have to use energy to move against an escalator?

24) Describe how cell membranes are used in active and passive transport.

25) How did the models you experienced today show how membranes function?

Cell Town

Directions:

Cell structures and functions are a lot like how a town is structured and functions. You need to draw a town and label each part of that town with a part of the cell that functions the same as that part of a town. This project can be done by hand or digitally. First, use the **internet** or your **textbook** to find each organelle's functions. Then find out what is in a town that would have the same functions. Write them down next to the organelles below. Then draw/build and label your town.

For example, the ***nucleus*** *is the cell's control center, so that would be like* ***City Hall****, so you would draw a* ***City Hall*** *and label it the* ***nucleus****.*

Cell wall

Cell Membrane

Nucleus – ***City Hall***

Nucleolus

Nuclear Envelope

Chromatin/DNA

Ribosomes

Smooth Endoplasmic Reticulum

Rough Endoplasmic Reticulum

Golgi Apparatus

Lysosome

Chloroplast

Mitochondria

Cytoplasm

Cytoskeleton

Protein Channels

Water Vacuole

Food Vacuole

Chromoplast

Leucoplast

Characteristics of Prokaryotic and Eukaryotic Cells

Directions:

You will need **prepared slides of bacteria**, **Amoeba**, **Paramecium**, and **Euglena,** or you can prepare wet-mount slides of these same organisms according to your teacher's instructions. You will also need the **internet**, your **textbook**, a **compound light microscope**, **lens wipes**, and have your teacher pull up videos on **YouTube** of these organisms interacting and feeding for you to see. **Looking at the materials and lab we will be using, what are the safety precautions we should take to protect ourselves and materials during the investigation?**

1) Focus each of these organisms under your microscope and draw a picture of each in the table below. Show your teacher you can safely and correctly focus these organisms by having your teacher come by and look at one of them centered and focused on your microscope. Teacher's initials:

Table 1

Organism	Picture	Describe movement & feeding
Bacteria (Prokaryote)		
Amoeba (Eukaryote)		

Paramecium (Eukaryote)		
Euglena (Eukaryote)		

2) Look in your textbook or on the internet and label each organism's important parts on the pictures you have drawn in Table 1.
3) Search and watch YouTube videos on how each of these organisms moves and feed and describe it in Table 1.

Questions:

1) Based on your observations, do the cells have the same shape? Explain.

2) Based on your observations, do the cells have the same size? Explain.

3) Based on your observations, do all cells have the same parts? Explain.

4) What cell structures do you see are common to all cells?

5) What cell structures are found only in eukaryotic cells?

6) Why do you think different cells have different shapes and sizes?

7) How do the eyespot and chloroplast work together to help the euglena survive?

8) What characteristics of an animal does a euglena possess?

9) What characteristic of a plant does a euglena possess?

10) How is the amoeba like a white blood cell that engulfs invading organisms in our bodies?

11) How are the pseudopods used in amoeba?

12) What are the functions of the cilia on the paramecium?

13) Which of these organisms seem to be the most advanced/complicated? Explain.

14) Which of these organisms would probably be the easiest to keep alive in class if you had a solar light source?

15) How big do you expect the chloroplasts and mitochondria to be if they are bacteria that act as organelles in eukaryotes?

16) From what you observed in this lab, how does this give evidence for endosymbiosis?

17) Should we call this concept an Endosymbiotic Hypothesis or Theory? Explain.

Draw a Detailed Picture of Bacteria

Directions:

1) Find a detailed picture of a bacterium on the **internet** or in your **textbook**, draw it, and label it below.

2) Write the function of each part of the Bacterium you labeled.

Draw a Detailed Picture of Paramecium

Directions:

1) Find a detailed picture of a Paramecium on the **internet** or in your **textbook**, draw it, and label it below.

2) Write the function of each part of the paramecium you labeled.

Paper Mates

Directions:

You will need two **pennies**. Heads represent a dominant trait, and tails represent a recessive trait. The parents are heterozygous for all four traits shown below. Flip two coins to see what each trait will be in each of the three offspring you create. Write the results in # 2. **Looking at the materials and lab we will be using, what are the safety precautions we should take to protect ourselves and materials during the investigation?**

Trait	Genotypes and Phenotypes
Eyes	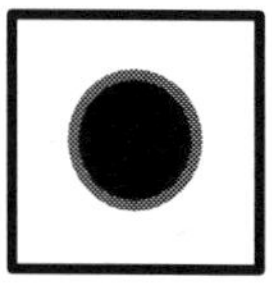 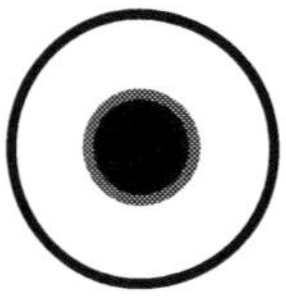 EE or Ee; ee
Nose	  NN or Nn; nn
Teeth	 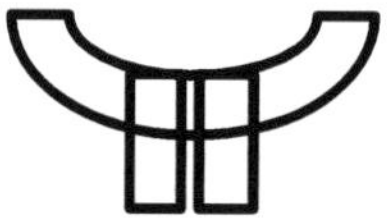 TT or Tt; tt
Hair	   HH; Hh; hh

1) **Parents (P Generation):** Draw the parents below. They are heterozygous for all traits.

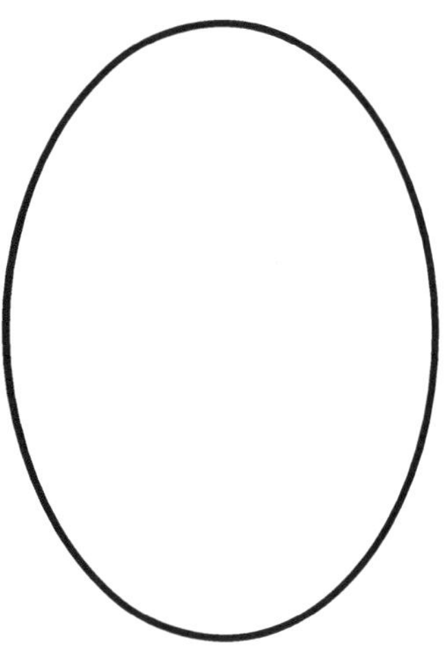
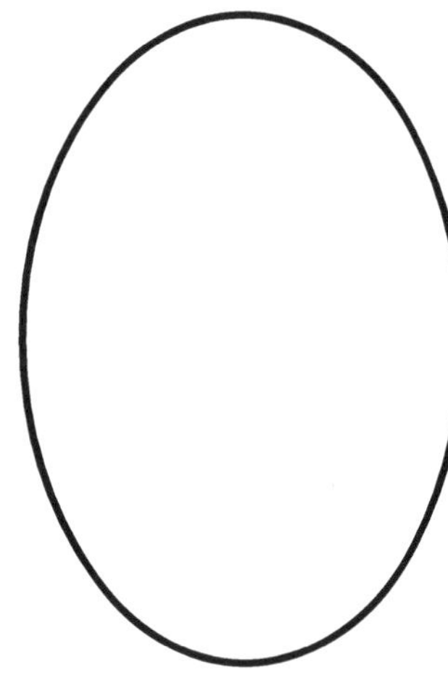

2) **Family (F1 Generation)**: Write down the genotypes of each trait that happened from flipping the coins for each offspring. Then draw the three offspring that came from the flipping of the coins.

Genotype: Genotype: Genotype:

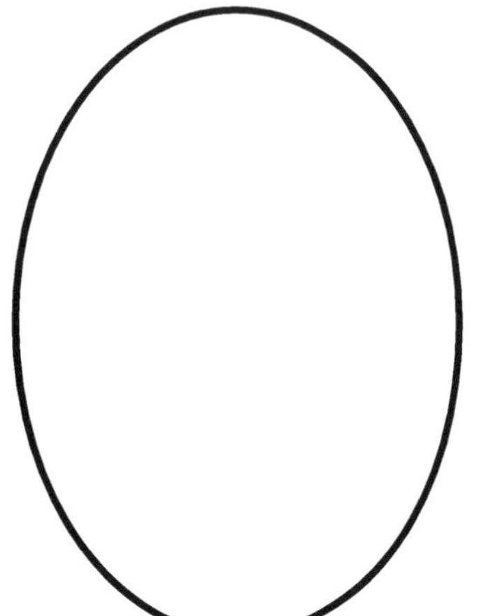
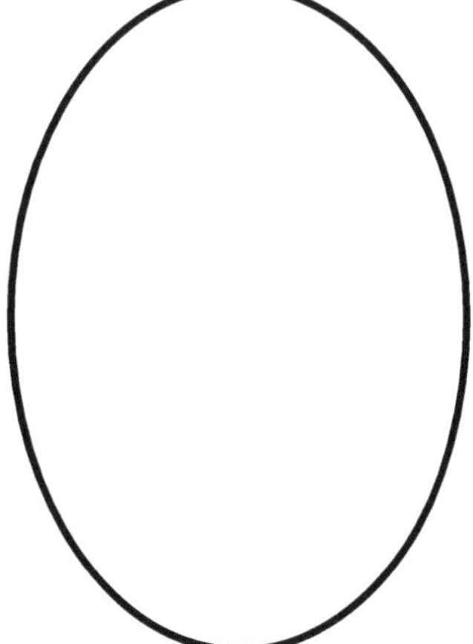
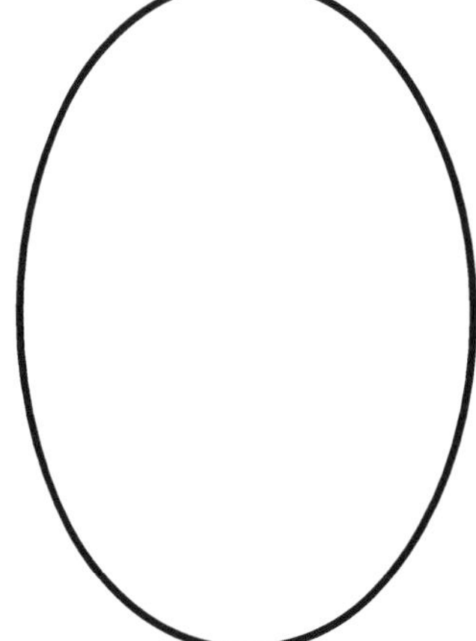

Create a Baby

Directions:

You will need one **penny** for each parent. One of you will be the mother, and one will be the father. **Looking at the materials and lab we will be using, what are the safety precautions we should take to protect ourselves and materials during the investigation?**

1) Begin with determining the sex of the baby. The mother can only pass on the female chromosome (X). The father can pass on a female (X) or a male chromosome (Y). Have the father flip the coin. Heads, the baby is male, tails, it is a female. Record this on Data Table 1.
2) Take turns flipping the coin to determine your baby's traits, starting with the head shape; heads will always be dominant, and tails will always be recessive. Fill in this information in Data Table 1. (**For example:** if the mother's coin lands on tails, she will pass on the recessive trait (r) for a round face. If the father's coin lands on heads, he will pass on the dominant trait (R) for an oval face. The baby's genotype will be (Rr), and its phenotype will be round-faced.) See page 236 for reference.
3) Repeat this information for each trait in Data Table 1. Use your own traits as the mother and father's genotype for eye color and hair color. (**For example:** if the mother has red hair, enter AAbb as the mother's genotype. If the father has dark brown hair, enter AaBB as the father's genotype. The mother would not have to flip since she can only give an Ab. The father would only flip for the "Aa"; the other gene will be B. So the baby will either be AABb or AaBb, depending on the flip.)
4) Draw your baby's face on page 237 based on the baby's phenotype on Data Table 1 and the way it looks on the Genotype/Phenotype Reference Sheet on page 236.

Data Table 1

Trait	Mother's Genotype	Mother's Coin Flip	Father's Genotype	Father's Coin Flip	Baby's Genotype	Baby's Phenotype
Gender	XX	X	XY		X	
Face Shape	Rr		Rr			
Chin Shape	Nn		Nn			
Freckles	Ff		Ff			

Dimples	Dd		Dd			
Lip Thickness	Tt		Tt			
Eye Brows	Bb		Bb			
Eyelash	Ll		Ll			
Ear Lobes	Ee		Ee			
Widow's Peak	Ww		Ww			
Hair Curliness	Cc		Cc			
Nose Size	Ss		Ss			
Hair Color						
Eye Color						

Questions:

1) Are all traits only dominant and recessive?

2) Did you have a boy or a girl?

3) How do hair and eye color genetics differ from the other genes in this activity?

4) How are nose size and hair curliness different from the other traits?

5) How is homozygous freckled different from heterozygous freckled?

Trait	Homozygous Dominant	Heterozygous	Homozygous Recessive
Face Shape	RR Round	Rr Round	rr Oval
Chin Shape	NN Noticeable	Nn Noticeable	Nn Less Noticeable
Freckles	FF Present	Ff Present	ff Absent
Dimples	DD Present	Dd Present	dd Absent
Lip Thickness	TT Thick	Tt Thick	Tt Thin
Eye Brows	BB Bushy	Bb Bushy	bb Fine
Eye Lashes	LL Long	Ll Long	Ll Short
Ear Lobes	EE Free	Ee Free	ee Free
Widow's Peak	WW Present	Ww Present	ww Absent
Hair Curliness	CC Curly	Cc Wavy	cc Strait
Nose Size	SS Small	Ss Medium	ss Large
Hair Color	AABB=Black AABb=Black AAbb=Red	AaBB=Brown AaBb=Brown Aabb=Blond	aaBB=Blond aaBb=Blond aabb=white (albino)
Eye Color	AABB=Deep Brown AABb=Deep Brown AAbb=Brown	AaBB=Green-Brown AaBb=Brown Aabb=Grey-Blue	aaBB=Green aaBb=Light Blue aabb=Pink (albino)

Modeling Cell Division

Directions:

You will need a set of **pipe cleaners**. Two pieces need to be cut to lengths of 1 inch, 2 inches, 3 inches, and 4 inches, respectively. **Looking at the materials and lab we will be using, what are the safety precautions we can take to protect ourselves and materials during the investigation?**

1) Each two pipe cleaners that are cut to the same length can be twisted together to make one chromosome with two chromatids. You should have for chromosomes total.
2) **Prophase**: Make a pile of all your chromosomes. There should be 4 of them. During this phase, the DNA winds up into chromosomes that can be seen with a microscope. How many chromosomes do you have in your pile?

3) **Metaphase**: Take your chromosomes and line them up with each other. How many chromosomes do you have now?

4) **Anaphase**: Separate your chromatids to opposite sides, so you have two lines of chromosomes. Each chromatid now becomes its own chromosome.

5) **Telophase I**: Now, make two piles out of your separated chromosomes. At the end of this phase, the chromosomes unwind, and two nuclei form. Cytokinesis separates the cell making two cells. How many chromosomes do you have in each cell?

6) Why do you think cells do this?

7) Where does this process happen?

Hand Models Showing Cell Division

Directions:

Follow the instructions for this model with your hands. You can do this together with your teacher in your class.

Mitosis:

1) Take both your hands and hold them together. Your fingers represent the chromosomes in **prophase**.
2) Line your fingers up next to each other, interlocking with the fingers of your left-hand touch the corresponding fingers with your right hand. Now you are modeling **metaphase** with the chromosomes lined up in a row.
3) Now separate your hands; as you do this, you are modeling **anaphase**.
4) Now make two fists. Each hand becomes a nucleus in two cells during **telophase**.

Cytokinesis:

1) Form a circle with both hands by putting your fingertips and thumbs together; this represents one cell.
2) Slowly bring your fingertips down to your thumbs, eventually bringing them together, pinching the circle into two circles; this shows how the cytoplasm of one cell separates into two cells during cell division.

Questions:

1) How do these models show the two processes in cell division?

2) How are these models not accurate?

Modeling Meiosis

Directions and Questions:

You will need two different color sets of **pipe cleaners**. Four pieces need to be cut to lengths of 1 inch, 2 inches, 3 inches, and 4 inches, two for each color. **Looking at the materials and lab we will be using, what are the safety precautions we can take to protect ourselves and materials during the investigation?**

1) Each two pipe cleaners that are cut to the same length and are the same color can be twisted together to make one chromosome with two chromatids.
2) Each pair of chromosomes (one of each color) represents a homologous pair of chromosomes.
3) **Prophase I**: Make a pile of all your chromosomes. There should be 8 of them. DNA has wound up into chromosomes. Four of one color and four of another. Each color represents the chromosomes given to you by each of your parents. This cell is **diploid (di-** meaning two**)**; it has double the chromosomes. How many chromosomes do you have in your pile?

    a. The first cell in males is called a **spermatocyte.**
    b. The first cell in a female is called an **oocyte.**

4) **Metaphase I**: Take your homologous pairs and line them up with each other, randomly placing different colors on opposite sides; this is how **independent assortment** occurs, lining homologous chromosomes up randomly.

5) **Anaphase I**: Separate your homologous pairs to opposite sides, so you have two lines of chromosomes.

6) **Telophase I**: Now, make piles out of your separated chromosomes. You have made **haploid** cells because they have half the chromosomes that they had before. At the end of this phase, the chromosomes unwind, and two nuclei form. Now we are halfway done. How many chromosomes do you have in each of your piles?

7) **Prophase II**: You already have Prophase II, two piles of chromosomes that wound back up.

8) **Metaphase II**: Take each of your piles and line up the chromosomes in them.

9) **Anaphase II**: Untwist each of the chromosomes, separating their chromatids. You should now have four lines of chromosomes.

10) **Telophase II**: Now, take your four lines of chromosomes and put them each into their own pile. These will end up being four different cells still being **haploid**. How many chromosomes are in each of your piles now?

    a. In males, all four will become **sperm** cells.
    b. In females, three of the cells will become small useless **polar bodies,** and one cell will become a larger **egg**.

11) **Fertilization**: This can be done by taking one of your four piles and combining them with another person's pile. This new cell would be called a **zygote**. This new cell/organism is now **diploid** because it has double the amount of chromosomes it had before. How many chromosomes do you have now?

Comparing Ratios of Monohybrid Crosses

Directions:

You will need a separate **sheet of paper** and two **pennies. Looking at the materials and lab we will be using, what are the safety precautions we can take to protect ourselves and materials during the investigation?**

1) We will say the penny's head's side is dominant for hair (H), and the tail's side is recessive for having no hair (h). On a separate sheet of paper, construct a Punnett square to find the expected genotype and phenotype ratios; write them in Data Table 1.
2) Create a tally sheet on that separate sheet of paper to mark how many HH, Hh, and hh happen when you toss/flip the coins. Put the totals in Data Table 1 for 10, 100, and 1000 tosses.

Data Table 1

# of Tosses	# of HH	# of Hh	# of hh hairless	Total hairy (HH+Hh)	Expected Genotype Ratio	Expected Phenotype Ratio	Experimental Genotypic Ratio	Experimental Phenotypic Ratio
10								
100								
1000								

Questions:

1) Which phonotype happened more, hairy or hairless? Explain why.

2) Which genotype appeared the most?

3) Why do you think this genotype appeared so often?

4) Which phenotype appeared the least?

5) Why do you think this phenotype appeared the least?

6) How did the expected ratios compare with the experimental ratios?

7) How many experimental tosses produced a ratio closest to the expected ratios?

8) Which type of ratio shows the probability that something will happen?

9) Which type of ratio shows what actually did happen?

10) Will the expected ratios always match the experimental? Explain why.

11) How do flipping coins model independent assortment?

12) Why is independent assortment a law?

Making a Karyotype

Directions:

1) Use the key below to build an individual's karyotype with a disorder. On page 245, you will need to label the remaining unlabeled chromosomes from 1-22, and X or Y. Pair #23 will be different if it is male, and the same bigger chromosome pair if female.
2) Next, use **scissors** to cut them out and either **tape** or **glue** them to the blank chart on page 247 at the appropriate spots. Make sure the centromeres are lined up on the horizontal line.
3) Identify where the abnormality is and research the **internet** or your **textbook** to identify the disorder.

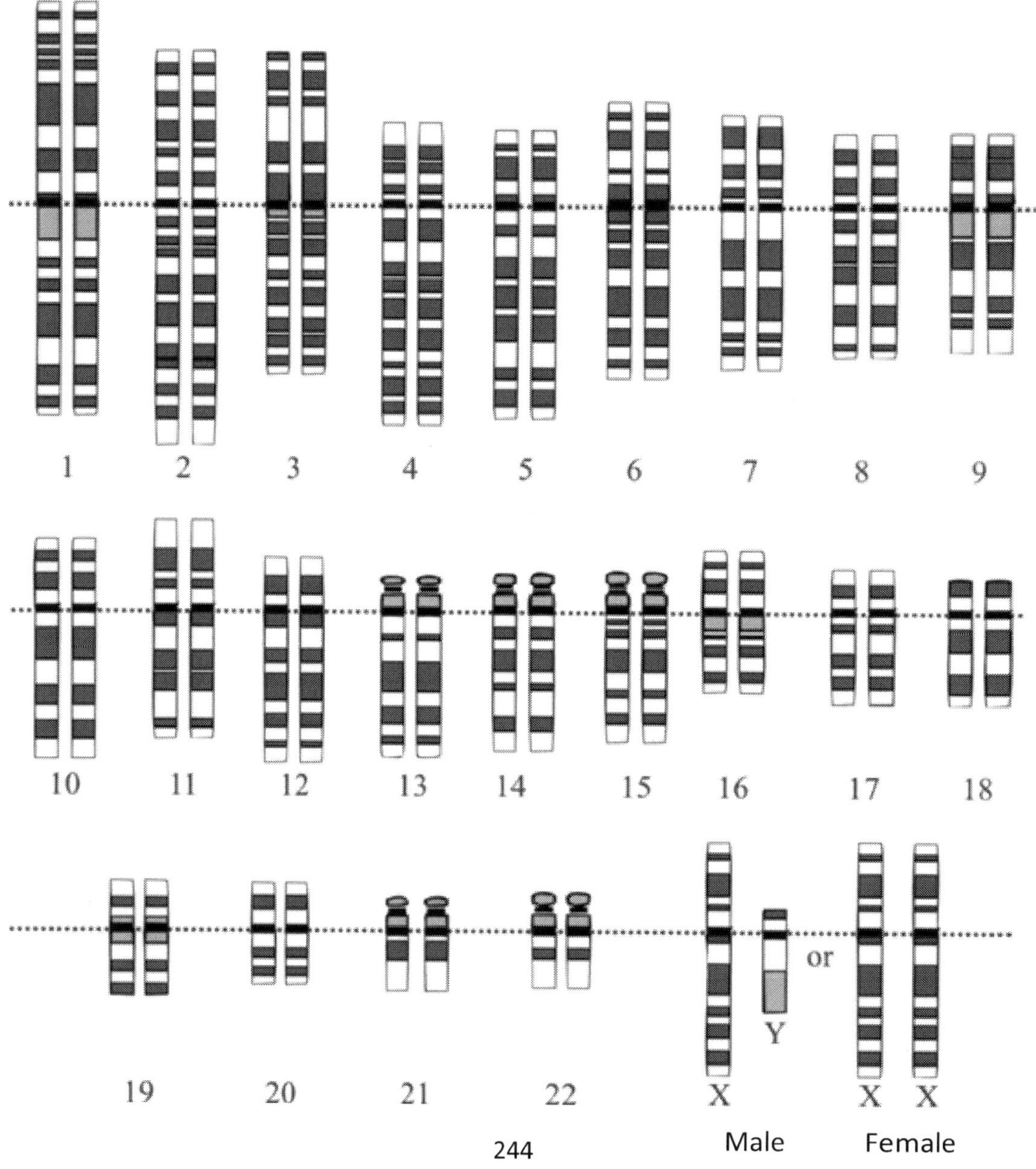

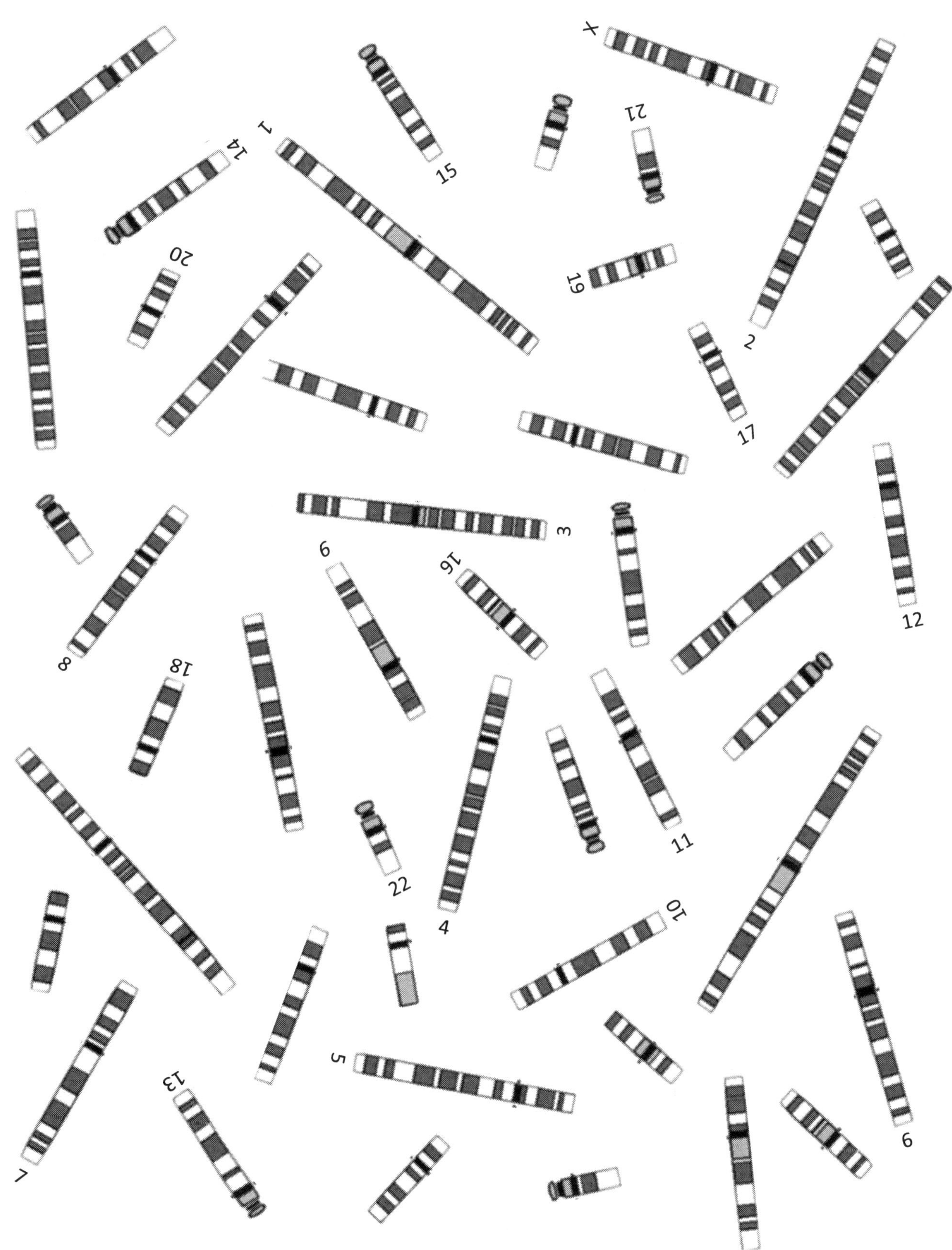
X
21
14
1
15
20
19
2
17
3
9
16
12
8
18
11
22
4
10
5
13
7
6

This page will be cut up from the page before!

Empty Chart

1 2 3 4 5

6 7 8 9 10 11 12

13 14 15 16 17 18

19 20 21 22 23

Questions:

1) How many sex chromosomes are in a normal human?

2) What is the difference between a normal male and a normal female on a karyotype?

3) Is your karyotype a male or female?

4) What did you observe was abnormal about your karyotype?

5) All the chromosomes that are not the sex chromosomes are the autosomes. How many autosomes were found in your karyotype?

6) How many autosomes are in a normal human karyotype?

7) What do you think would be worse, too many chromosomes or missing chromosomes? Explain.

8) How could you tell the different chromosomes apart?

9) How could you tell which chromosomes are homologous when building the karyotype?

10) Which phase of mitosis do you think was used to get the image of the chromosomes? Explain.

11) How do karyotypes show genes in chromosomes determine inherited traits?

12) What kind of impact can this type of technology have on human lives and society?

13) What is the abnormality called that you found in your karyotype?

Lego Mutation Models

Directions:

You will need a lot of **4 block Legos** of at least three different colors or a bag of **Mega Blocks** to build chromosome models simulating some of the different types of mutations. **Looking at the materials and lab we will be using, what are the safety precautions we can take to protect ourselves and materials during the investigation?**

1) Each chromosome is a pattern of information telling how to put together an organism. Use the images below and follow your teacher's instructions to stack your blocks with a pattern with two different colors to simulate a chromosome you will start with.
2) Produce a deletion mutation by taking a section out of your chromosome and connecting it back together.
3) Produce a duplication mutation by adding a pattern already in the chromosome.
4) Produce an inversion mutation by reversing the order of a section in your stack.
5) When simulating an insertion, use a third color to make another stack so you can see the insertion genes.
6) When simulating the translocation mutation, make two stacks where the pieces of each chromosome swap places

Types of Mutations

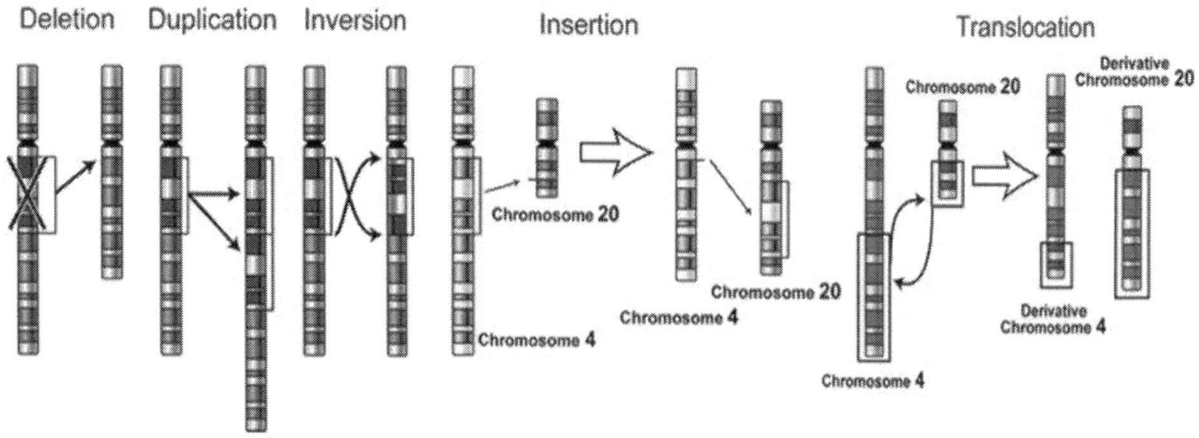

Questions:

1) What do you think could happen if you delete information from an organism?

2) What do you think could happen if you duplicate information in an organism?

3) What do you think could happen if you change the order of the information in an organism?

4) What do you think could happen if you add information to an organism?

5) How could each of these types of mutations benefit an organism?

   a. Deletion

   b. Duplication

   c. Inversion

   d. Insertion

   e. Translocation

6) How could each of these types of mutations harm an organism?

    a. Deletion

    b. Duplication

    c. Inversion

    d. Insertion

    e. Translocation

7) Which seems to be the most dangerous type of mutation you simulated today? Explain why.

8) How does mutation help the process of Natural Selection?

Protein Synthesis of the Quaddie

Directions:

You will need **colored pencils** to draw the Martians we did not recently find on Mars. Looking at their genetic code, we can see their weird protein structure that acts differently on Mars than on Earth. We suspect that some large collision or explosion shot microscopic organisms out of the Earth at some point in its past, sending them to Mars. Mutations that occurred along with the different force of gravity and type of atmosphere caused this organism to develop differently and evolve to live in the conditions of Mars. Strangely mutations have taken place to cause the genes to shorten themselves to four codons long, which is extremely short compared to all organisms on Earth. Because these organisms only have eight genes, each with four codons, we call them Quaddies. (Remember code: **DNA**: **A-T**, **C-G**; **RNA**: **A-U**, **C-G**)

1) Use the DNA code in Data Table 1 for each gene to configure the mRNA codons. Then use those mRNA codons to see which tRNA would connect to the amino acids.
2) Then use the mRNA again with the Codon Table on the top of page 255 to find the amino acid sequence.
3) Once you have the amino acid sequences, you can use these to code for the organism's traits in Data Table 2 to fill in Data Table 1.
4) Use these traits to draw what you think this Quaddie looks like in Picture 1 on the bottom of page 255. Make sure you do not draw things that do not have a code.

Data Table 1

Gene 1	**Gene 2**
DNA: TAC AGA CTT CTG mRNA: *AUG* ______ tRNA: *UAC* ______ Amino Acids: *Methionine* ______ ______ Trait: ______	DNA: GTA ATG TTT GGA mRNA: ______ tRNA: ______ Amino Acids: ______ ______ Trait: ______
Gene 3	**Gene 4**
DNA: CAT GCC TCA CCC mRNA: ______ tRNA: ______ Amino Acids: ______ ______ Trait: ______	DNA: CGA GCT AAA TGA mRNA: ______ tRNA: ______ Amino Acids: ______ ______ Trait: ______

Gene 5	Gene 6
DNA: TAA CGT ATA GCT mRNA: ______ tRNA: ______ Amino Acids: ______ ______ Trait: ______	DNA: CAC TGC TTA AGG mRNA: ______ tRNA: ______ Amino Acids: ______ ______ Trait: ______
Gene 7 DNA: ACC GTC ACA CCA mRNA: ______ tRNA: ______ Amino Acids: ______ ______ Trait: ______	**Gene 8** DNA: AAG TTG CTC ATC mRNA: ______ tRNA: ______ Amino Acids: ______ ______ Trait: ______

Data Table 2

Amino Acid Sequence	Trait
Isoleucine-Alanine-Tyrosine-Arginine	Eyespot
Isoleucine-Alanine-Phenylalanine-Threonine	No eyespot
Aspartic Acid-Arginine-Serine-Cysteine	Hairy
Aspartic Acid-Arginine-Serine-Glycine	No hair
Valine-Threonine-Asparagine-Serine	Fantail
Histidine-Threonine-Asparagine-Serine	Skinny tail
Methionine-Serine-Glutamine-Aspartic Acid	4 legs
Methionine-Serine-Glutamic Acid-Aspartic Acid	8 legs
Phenylalanine-Asparagine-Glutamic Acid-Stop	4 nostrils
Phenylalanine-Alanine-Glutamic Acid-Stop	2 nostrils
Alanine-Leucine-Phenylalanine-Tryptophan	No antennae
Alanine-Leucine-Phenylalanine-Threonine	4 antennae
Histidine-Tyrosine-Lysine-Proline	Green skin
Proline-Tyrosine-Lysine-Histidine	Red skin
Glycine-Glutamine-Cysteine-Tryptophan	Green spots
Tryptophan-Glutamine-Cysteine-Glycine	Red spots

Codon Table

First Base	U	C	A	G	Third Base
U	Phenylalanine	Serine	Tyrosine	Cysteine	U
	Phenylalanine	Serine	Tyrosine	Cysteine	C
	Leucine	Serine	Stop	Stop	A
	Leucine	Serine	Stop	Tryptophan	G
C	Leucine	Proline	Histidine	Arginine	U
	Leucine	Proline	Histidine	Arginine	C
	Leucine	Proline	Glutamine	Arginine	A
	Leucine	Proline	Glutamine	Arginine	G
A	Isoleucine	Threonine	Asparagine	Serine	U
	Isoleucine	Threonine	Asparagine	Serine	C
	Isoleucine	Threonine	Lysine	Arginine	A
	Methionine	Threonine	Lysine	Arginine	G
G	Valine	Alanine	Aspartic acid	Glycine	U
	Valine	Alanine	Aspartic acid	Glycine	C
	Valine	Alanine	Glutamic acid	Glycine	A
	Valine	Alanine	Glutamic acid	Glycine	G

Second Base

Picture 1

Questions:

1) What is the difference between the DNA codons and mRNA codons?

2) What do you think is the amino acid Methionine's function if it is always used at the beginning of the gene code in real life?

3) What is the function of the Stop codon?

4) How is this organism genetically different from those found on Earth?

5) What part of this activity is a part of transcription?

6) What part of this activity is part of translation?

7) How are mRNA codons different from tRNA anticodons?

8) What if we know the amino acid sequence of a gene? How could we find the DNA code for that same gene?

9) Is there only one DNA code for that one gene? Or could multiple DNA codes code for the same amino acid sequence? Explain.

10) What has to change for a protein to change, the DNA or amino acid? Explain.

11) How is this model different from real life?

Protein Synthesis Role Play

Directions:

You will need a **box** (to be the nucleus), two **twisted phone cords** (to be the DNA molecule), **masking tape** (with a series of codons written on it to act as mRNA), **connecting alphabet baby letters** or **shapes** (to be the amino acids), **students** will be the tRNA (that bring the amino acids to the ribosome), a **plastic Walmart bag**, and your **teacher** and a **chair** (will act as the ribosome). **Looking at the materials and lab we will be using, what are the safety precautions we should take to protect ourselves and materials during the investigation?**

Construction Pre-Lab:

1) **Nucleus:** On the box, label the nucleus for everyone to see.
2) **DNA:** take two twisted phone cords and push them together to build the two strands of spiral DNA. Put the DNA in the nucleus box.
3) **mRNA:** On the masking tape, write a series of mRNA codons (starting with **AUG**) that do not repeat themselves but are in a random order for as many baby letters as you have to represent the 64 available codons. Carefully place another strand of tape on the back to cover the sticky back of the tape up. Place the mRNA inside the nucleus box.
4) **Amino Acids:** On the connecting baby toys that can build a long chain, write the complementary tRNA anticodons on each piece that complement the mRNA codons.
5) **tRNA:** Randomly distribute all the baby toys with the tRNA anticodon on them to all the students in the class. Try to have everyone get the same amount.
6) **Ribosome:** place a chair for the teacher to sit in the middle of the room.

Modeling:

7) **Transcription:** The teacher will go to the nucleus box, unwind some of the phone cord DNA, and pull the masking tape mRNA from where the DNA splits to show transcription occurs there.
8) Pull all the mRNA tape out of the nucleus and put the DNA back together. Transcription is now over.
9) The teacher will now take the mRNA tape and sit in the chair to put the ribosome together. The teacher is one half of the ribosome, and the chair is the other.
10) **Translation:** The ribosome teacher sitting in the chair (the teacher cannot get up from the chair in the model) now will read the first codon to the class. The student (tRNA) who has the complementary anticodon on their amino acid toy piece will get up and

take that piece to the teacher (ribosome). The teacher (ribosome) will inspect the amino acid anticodon to ensure it is correct. If it is correct, the teacher will link the amino acid to the chain. The ribosome will reject it by returning it to the student transfer RNA if it is incorrect.

11) The procedure in # 10 will be repeated until the entire mRNA chain is read and all the amino acids are put into the protein chain.
12) Once the chain is completed, the teacher folds up the chain to show how it makes a complicated shape. Put the protein chain in the plastic bag (packaging it) to send away to do its function. If the chain breaks on accident, we can say that the protein was denatured by something and will not work properly for its function; it might also do something else.

Questions:

1) How did transcription take place?

2) What represented the DNA?

3) What represented the nucleus?

4) What represented the mRNA?

5) How did translation take place?

6) What represented the amino acid?

7) What represented the tRNA?

8) What represented the ribosome?

9) What was the function of the tRNA?

10) What were the functions of the ribosome?

11) What happened to the protein chain at the end?

12) Describe how protein synthesis shows us how genes in chromosomes determine inherited traits in organisms.

Models of Macromolecules

Directions:

You will need a **molecular model kit**, a **Periodic Table**, and your **textbook** or **internet**. **Looking at the materials and lab we will be using, what are the safety precautions we should take to protect ourselves and materials during the investigation?**

1) At the top of your periodic table, label it like this just below:

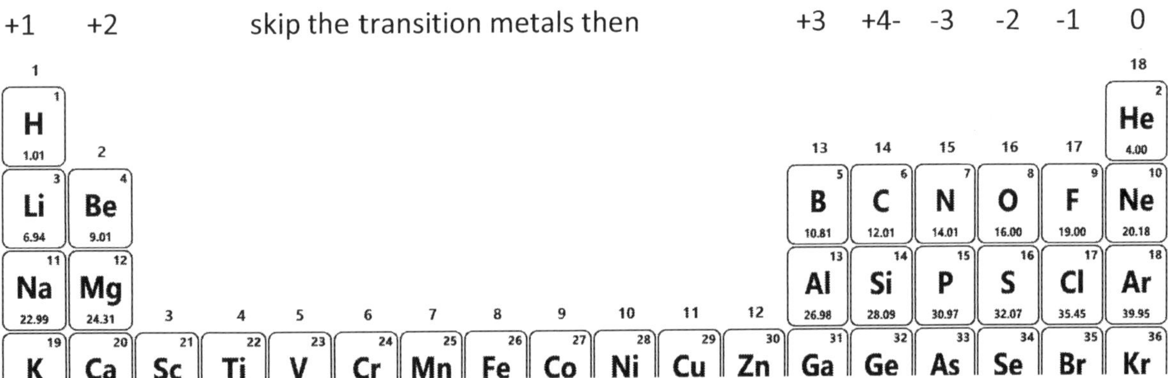

2) Different kits have different colors. In my kit, the:

   a. +1 (one-prong white) represents the Alkali Metals
   b. +2 (two-prong yellow) represents the Alkaline Earth Metals
   c. +3 (three-prong blue) represents the Boron Group
   d. +/- 4 (four prong black) represents the Carbon Group
   e. -3 (three-prong red) represents the Nitrogen Group
   f. -2 (two-prong blue) represents the Oxygen Group
   g. -1 (one-prong green) represents the Halogens
   h. The white tube is the bond

3) The different pieces in #2 represent the elements in those groups. Use pictures of large molecule monomers from your textbook or the internet to help you put them together. Then follow the directions to put amino acids together to make proteins.

   a. **<u>Protein</u>**: **Amino acids** are the **monomers** put together with **peptide bonds,** the building blocks of cell parts, enzymes, and hormones. An average usable protein molecule is 300-400 amino acids long.

Questions:

1) From what you have observed in this lab, what are the monomers?

2) What are the polymers?

3) How many atoms are in the monomer you made in this lab?

4) Can you count how big the polymer would be?

5) If we were to build a cell with these model pieces, how big do you think it would be?

6) Is the chemical organization of life simple or complex? Explain.

Virtual Investigations that go with Structure and Function of Life Units

ExploreLearning.com

Cell Structure

Cell Types

Paramecium Homeostasis

Diffusion

Osmosis

Osmosis STEM Case

Osmosis Handbook

Cell Division

Meiosis

Meowsis STEM Case

Meowsis Handbook

Building DNA

Inheritance

Chicken Genetics

Mouse Genetics (One Trait)

Mouse Genetics (Two Traits)

Fast Plants – 1 Growth and Genetics

Fast Plants – 2 Mystery Parent

Human Karyotyping

RNA and Protein Synthesis

Dehydration Synthesis

Heredity and Traits STEM Case

Heredity and Traits Handbook

Phet.colorado.edu

Gene Expression Essential

Unit 11: Ecological Changes

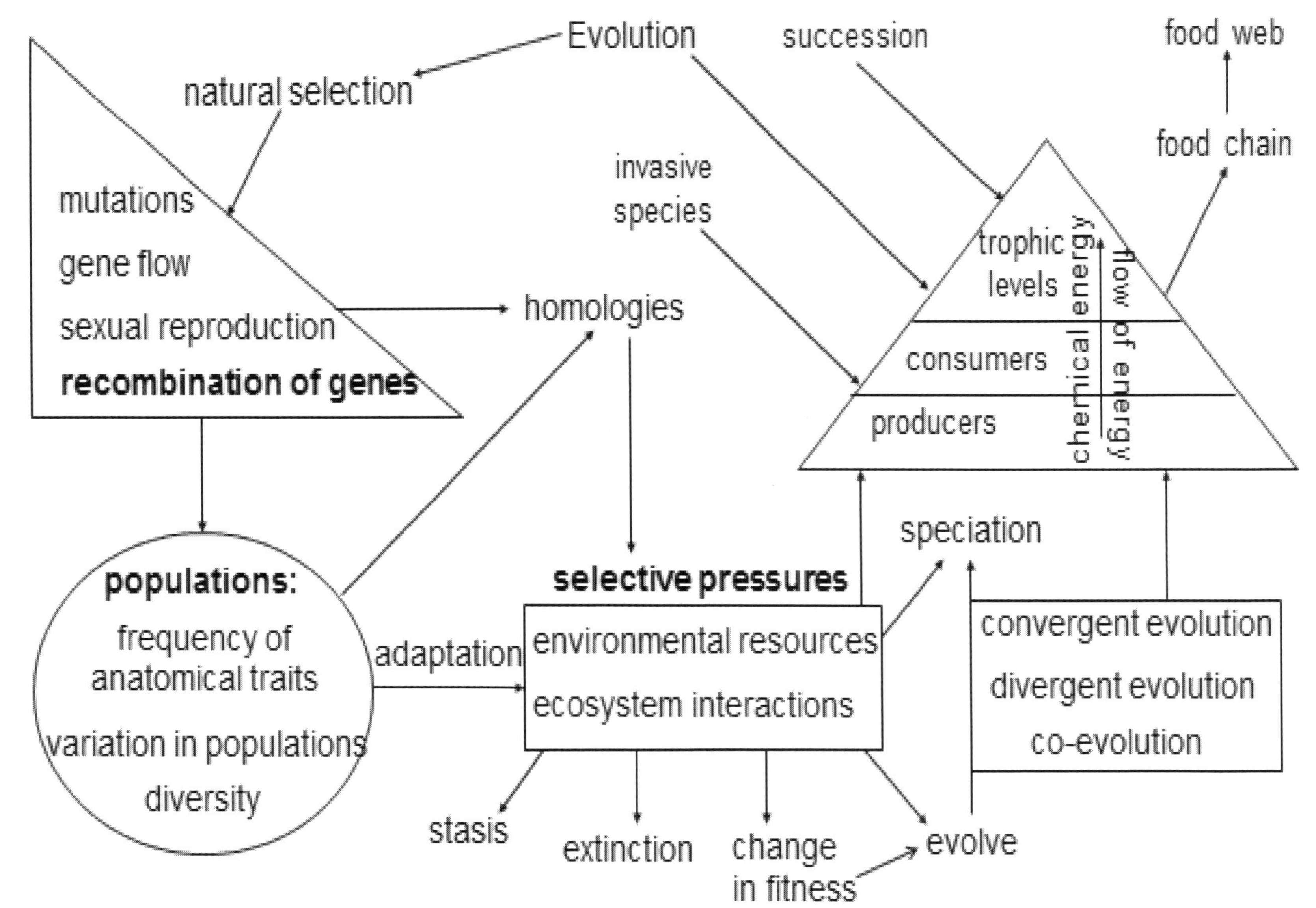

Unit 11: Ecological Changes

Succession			
Primary Succession	Both	Secondary Succession	
Caused by	Happen by	Caused by	
Nature	Nature	Nature	Human influence
volcanic eruption	pioneer community	natural disasters	mining
glaciers	change	hurricanes	fires
Land:	habitats	flood	deforestation
Aquatic:	populations	fires & deforestation	municipal development
	communities	mudslides avalanche	population growth
	climax community	tsunami	
starts on rock	feedback loops	starts on soil	

regional changes → environment → global effect

Variation Within a Population

Directions:

Each student will need ten **leaves** of any kind (must all be of the same species), ten **shelled nuts** or **seeds** (all of the same species) of any kind, and a **metric ruler**. The more students you have, the bigger your data set and the better results you will get. **Looking at the materials and lab we will be using, what are the safety precautions we should take to protect ourselves and materials during the investigation?**

1) Go out to a tree and randomly take ten leaves off the tree. When you take the leaf off, make sure you do not rip any part of the leaf. Measure the length of the longest part of the leaf in millimeters and write this data in Data Table 1.
2) Randomly take ten whole shelled nuts or seeds out of the bag. Do not use the ones that are broken. Measure the length in millimeters and write this data in Data Table 1.
3) Find the measurement of the shortest leaf in the class and the longest leaf in the class. Then fill in the equal increments between those measurements to make 14 different groups. Then take a class count of how many leaves fit in each of those categories. Write this data in Data Table 2.
4) Find the shortest nut pod/seed measurement and the longest nut pod/seed in the class. Then fill in the equal increments between those measurements to make 14 different groups. Then take a class count of how many nut pods fit in each of those categories. Write this data in Data Table 3.
5) Measure your pinky length from the crevice next to it (without stretching the skin webbing between your pinky and your ring finger) to the tip. Do not count the fingernail. You might want to total your data for the whole day to have enough numbers to show good data. Have your teacher keep your measurement on their roster. What is the length of your pinky?

6) Find the measurement of the shortest pinky of all the classes and the longest pinky of all the classes. Then fill in the equal increments between those measurements to make up 13 different groups. Then take a class count of how many pinkies fit in each of those categories. Write this data in Data Table 4. Then fill in the pinky data and let the students copy the all-day count the next day.

7) Make a graph of the class counts of tree leaf data on Graph 1.
8) Make a graph of nut pod/seed length data on Graph 2.
9) Make a graph of the pinky data for the whole day for Graph 3.

Data Table 1

	1	2	3	4	5	6	7	8	9	10
Leaf Length (mm)										
Nut Pod length (mm)										

Data Table 2

Leaf Length (mm)														
Class Count														

Data Table 3

Nut Pod Length (mm)														
Class Count														

Data Table 4

Length of Pinky (mm)													
Class Count													
All-day Count													

Graph 1

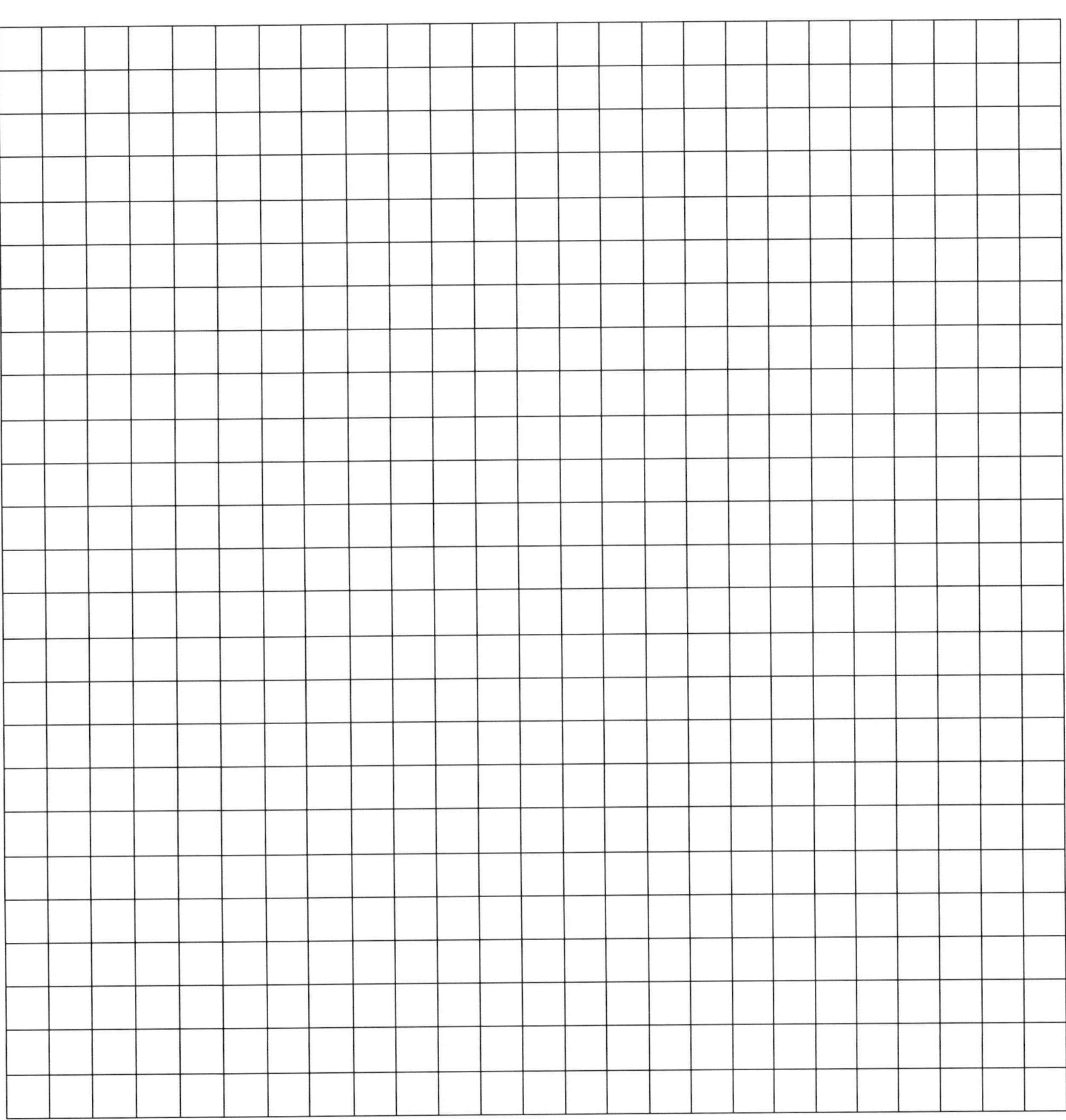

Graph 2

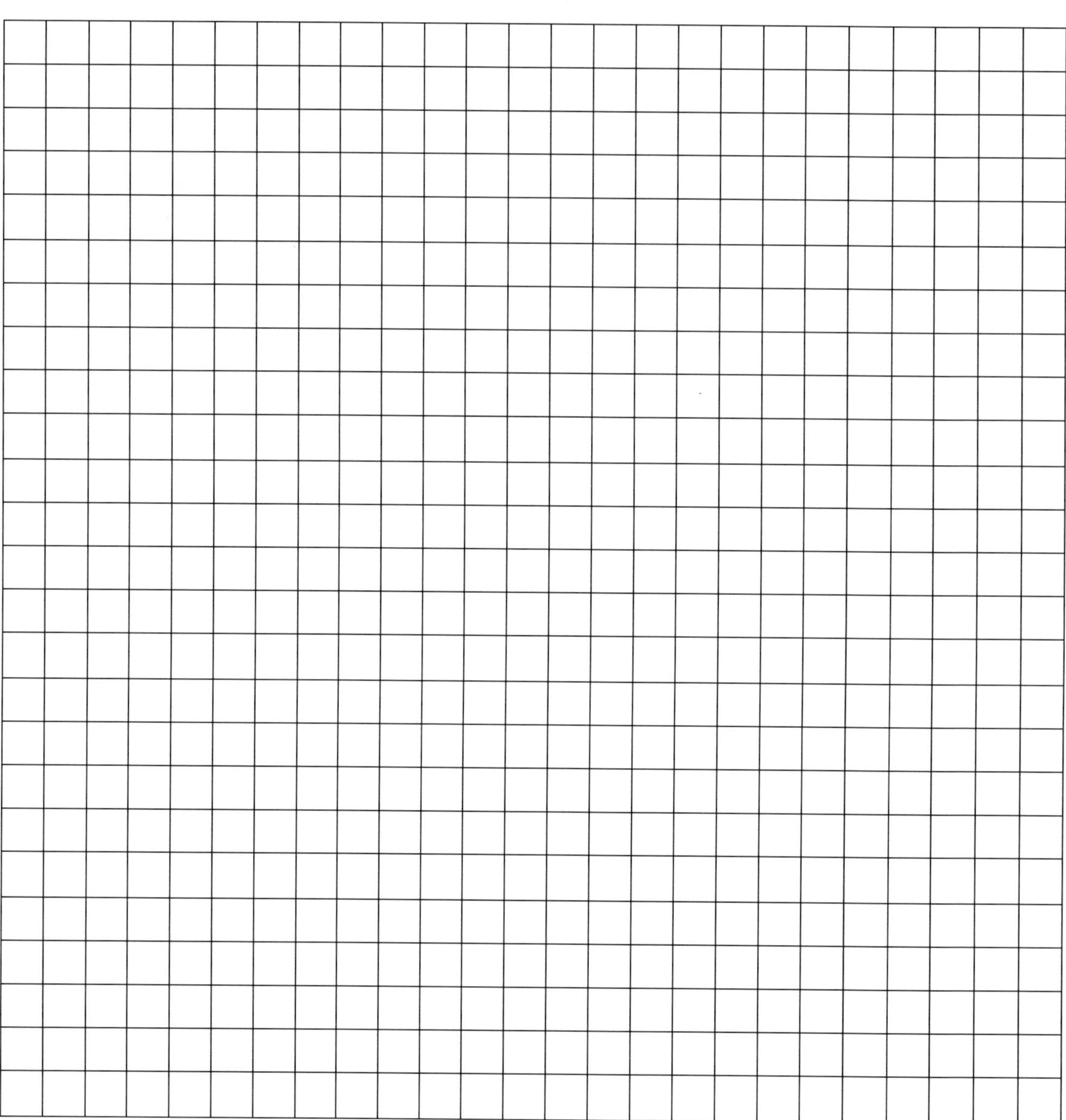

Graph 3

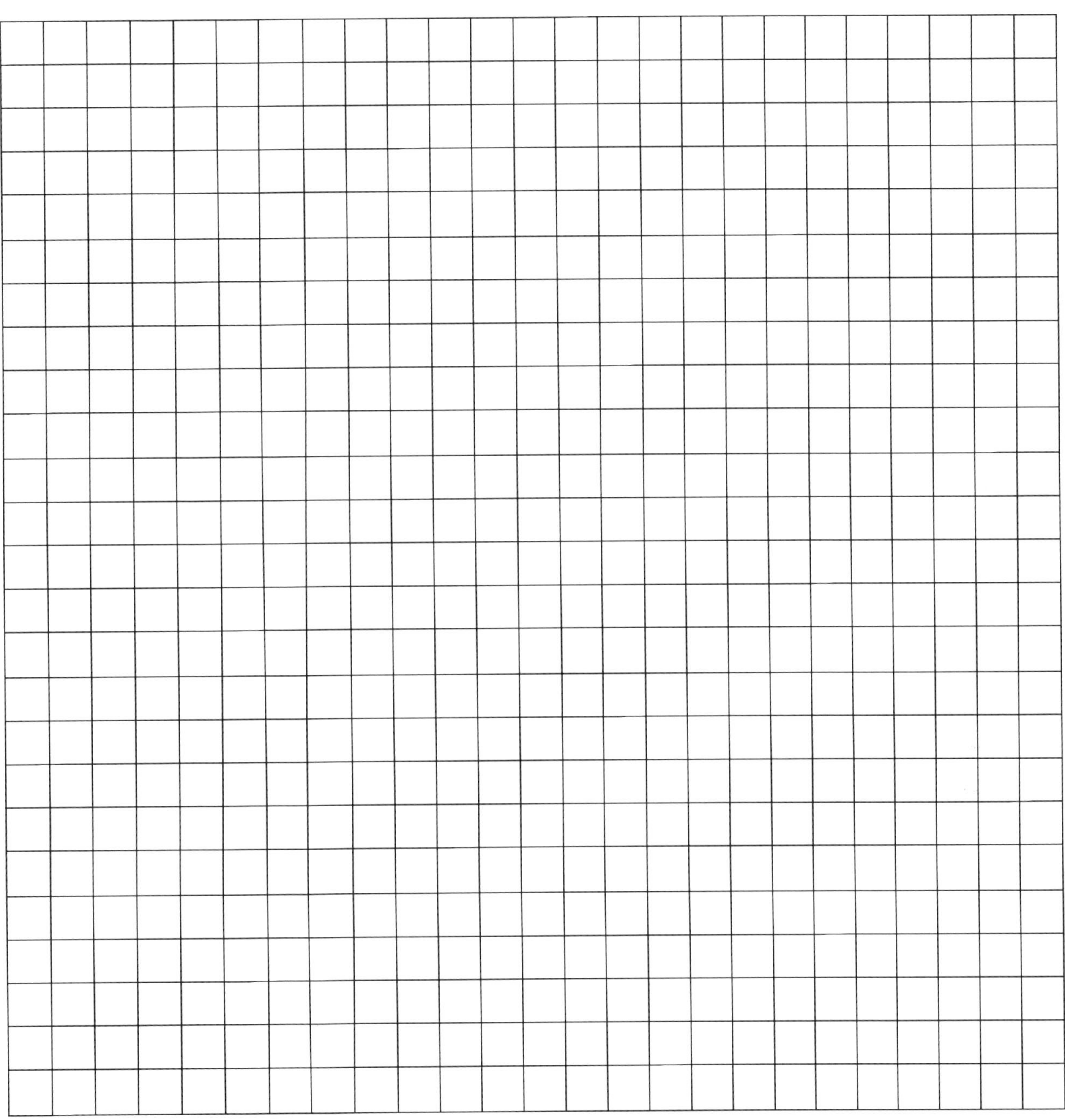

Questions:

1) What was the shape of each of those graphs?

2) Were all the measurements the same for all the leaves, seeds, and pinkies?

3) Why do you think populations have variation?

4) How can this be an advantage?

5) In natural selection, a certain trait is selected by nature to fit best in the environment. How can having a variety help a population survive when there is a sudden change in the environment?

6) Will the population look the same 1000 years after the environment's change selected out new successful individuals? (Explain)

7) If the population changes even slightly, evolution takes place. Does evolution take place if an individual dies? (Explain)

8) Does evolution take place if an individual is born? (Explain)

9) Where does evolution happen (to the individual or a population)?

10) How could big leaves become an advantage?

11) How could big leaves become a disadvantage?

12) How could long toes become an advantage?

13) How could long toes be a disadvantage?

14) How does variation set up a population for speciation?

15) What could be some sources of error in this investigation?

Goldfish Evolution

Directions:

You will need **food serving gloves** for the teacher, a **large mixing bowl**, **paper plates**, **cheese-flavored Goldfish Crackers**, and **pretzel flavored Goldfish Crackers**. **Looking at the materials and lab we will be using, what are the safety precautions we should take to protect ourselves and materials during the investigation?**

In this activity, students will represent predators, a goldfish-eating shark, which selectively preys upon goldfish in small populations. This shark likes to eat two kinds of fish: **yellow fish (cheese-flavored)** and **brown fish (pretzel flavored)**. The yellow fish are easy for you to see, so they are easy to catch and eat. Brown fish travel more quickly and can evade capture more easily. Because of this, you eat only yellow fish, unless there are no yellow fish around, in which case you eat the brown fish. Fish are replaced with individuals randomly selected from an ocean (mixing bowl full of Goldfish crackers). Brown fish is determined by the presence of a dominant allele (B), and yellow fish by a recessive allele (b).

1) Send one student from your group with a paper plate to collect a random population of 10 fish (crackers) from the mixing bowl (ocean). Your teacher will place them on your plate for you.
2) In data table 1, for generation 1, record the number of yellow and brown fish in the population.
3) Choose three yellow fish from the population and eat them. If you do not have any yellow fish, fill in the missing number by eating the brown fish for a total of 3 fish eaten.
4) Send one student to the bowl (ocean) to get three more random fish and add them to your population.
5) In data table 1, for generation 2, record the number of yellow and brown fish.
6) Repeat steps 3-5 until you have data for all five generations.

Data Table 1

Generations	#of Gold Fish	# of Brown Fish	% of Gold Fish	% of Brown Fish
1				
2				
3				
4				
5				

7) Using the information from Data Table 1 on page 274, plot your data on the graph below to show how your population changed over time. For each generation, plot two separated bars: use one color to represent the percent population of goldfish, and use a different color to plot the percent population of brown fish.

Graph 1

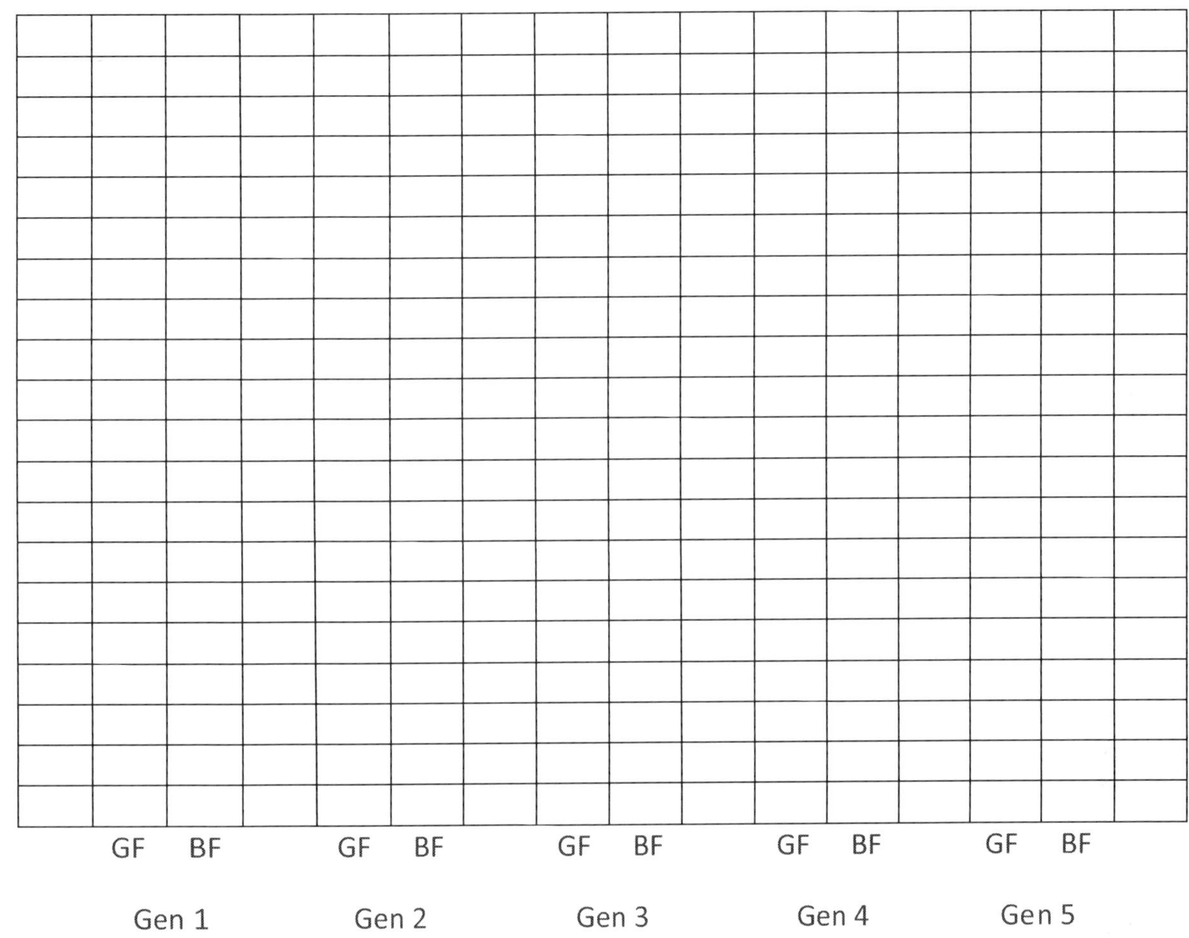

Questions:

1) How did the number of yellow fish change from generation 1 to 5?

2) Which phenotype was reduced in this population over time? Why?

3) What event occurs if there is a change in a population over time?

4) Explain what would happen over time if the brown fish were easier to catch?

5) What would happen if both fish were equally easy to catch?

6) How does this model show the Theory of Evolution and how speciation can occur?

Changing Environments for Beads

Directions:

You will need **red**, **white**, and **blue beads** in a **bowl, red, white,** and **blue construction paper**, and **colored pencils. Looking at the materials and lab we will be using, what are the safety precautions we should take to protect ourselves and materials during the investigation?**

1) Have each group randomly get 10 beads out of the bowl and place them on white paper. The beads will represent a population with three variants in them, and the paper will represent the environment they are in. Count how many beads there are of each color and write this in Data Table 1 for the first generation.
2) The students will represent a predator of the beads. Have the students in each group take out three beads that do not match the environment's background, place them back into the bowl, and randomly pick three more beads out of the bowl (if you do not have any that don't match the background take ones that do match until you have three).
3) Add the three new beads to the paper, count how many beads there are for each color in the population, and write this down in Data Table 1 for Generation 2.
4) Repeat steps 2 and 3 for six more generations.
5) After completing eight generations change the white background to red or blue and predict how your population will change over time.
6) Repeat steps 2 and 3 for eight generations and write this data in Data Table 2.
7) Once you have completed both Data Tables, graph your data for Data Table 1 on Graph 1 and Data Table 2 on Graph 2, making a line graph using red (for red beads), black (for white beads), and blue (for blue beads) colored pencils.
8) Once the graphs are completed, answer the questions that follow.

Data Table 1

Bead Color	Gen 1	Gen 2	Gens 3	Gen 4	Gen 5	Gen 6	Gen 7	Gen 8
Red								
White								
Blue								

Data Table 2

Bead Color	Gen 1	Gen 2	Gen 3	Gen 4	Gen 5	Gen 6	Gen 7	Gen 8
Red								
White								
Blue								

Graph 1 (Line Graph) Color of Background: White

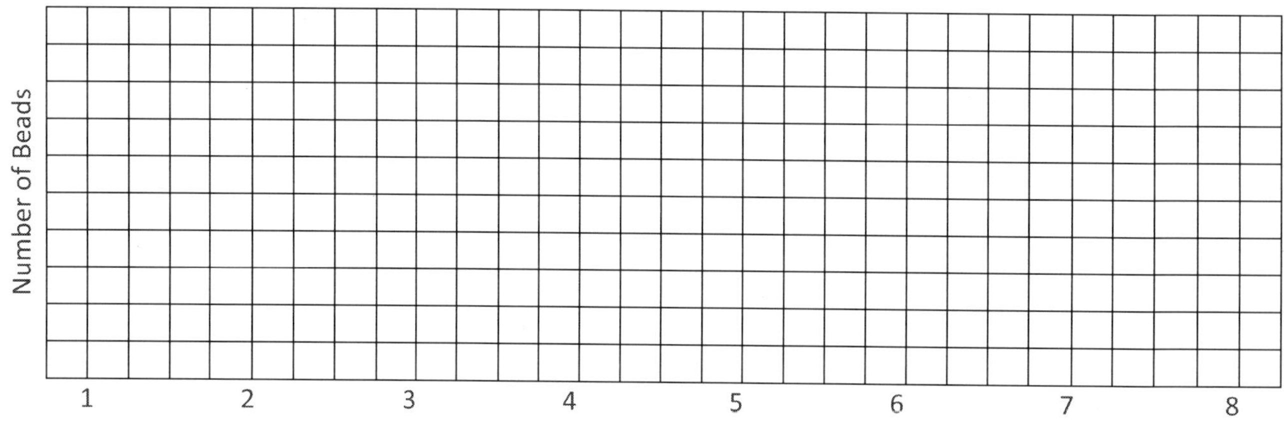

Graph 2 (Line Graph) Color of Background: ______________________

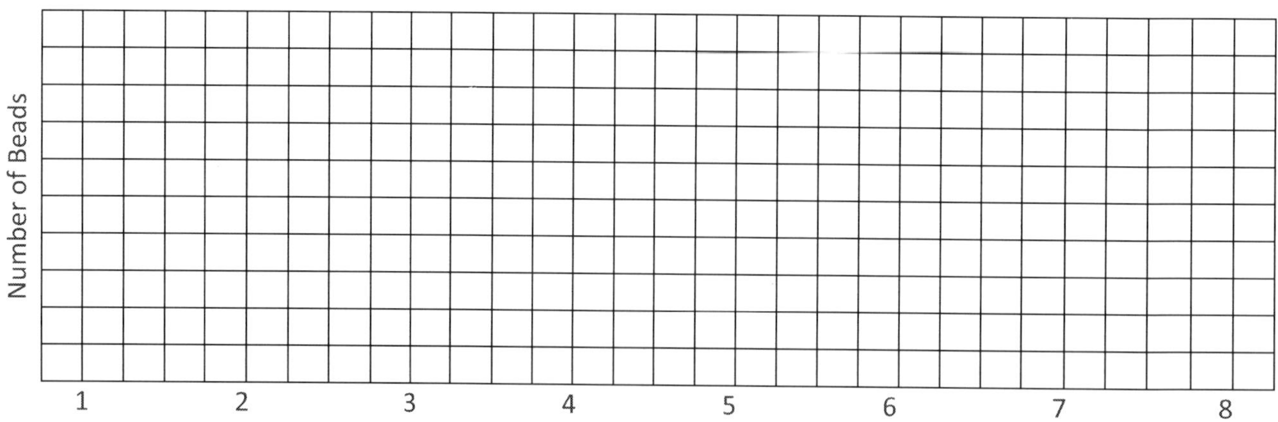

Questions:

1) How did the population of beads change when there was a white background?

    a. Why would this be useful in nature?

2) How did the population of beads change when there was a red or blue background?

3) Why would variations in a population benefit a population when an environment changes?

4) How could having variations in a population be hazardous to some individuals in the population?

5) What could happen to this population if the background turned green?

6) Which population would be more fit, one with little variation or one with lots of variation? Explain why.

7) How was this a good model for showing how variations within the population help populations adapt to changing environments?

8) How was this model not accurate?

Causes of Invasive Species

Directions:

Use the **internet** and your **textbook** to research invasive species and answer the following questions.

1) Give three examples of how climate change is causing invasive species.
   a.
   b.
   c.
2) Give three examples of how humans introduce invasive species.
   a.
   b.
   c.
3) Give three examples of when humans are the invasive species.
   a.
   b.
   c.

Humans Changing Ecosystems

Directions:

Use the **internet** to research how ecosystems changed when humans added or removed organisms from an ecosystem. Include which organisms were affected and how the food webs changed.

1) Humans moved to an uninhabited island, Mauritius, east of Madagascar, and brought pets.

2) Humans move into a new area and clear the land to build houses.

3) Humans release Burmese pythons in the Florida Everglades.

4) Humans release Asian carp into the Mississippi river to eat plants covering the river's bottom.

5) Humans plant crops on unused land.

6) Nutria were brought from South America to North America for the fur trade.

7) Wild boars were brought to the US for hunting and food.

8) Domesticated cats were bred and then released into the environment.

9) Humans chop down trees in the rainforest to make a field to raise cattle in Brazil.

10) In 1890 and 1891, European starlings were introduced to Central Park in New York City.

What Happens to the Food Web?

Directions:

Use the food web below to predict what would happen to the populations of the organisms when a change happens to the ecosystem.

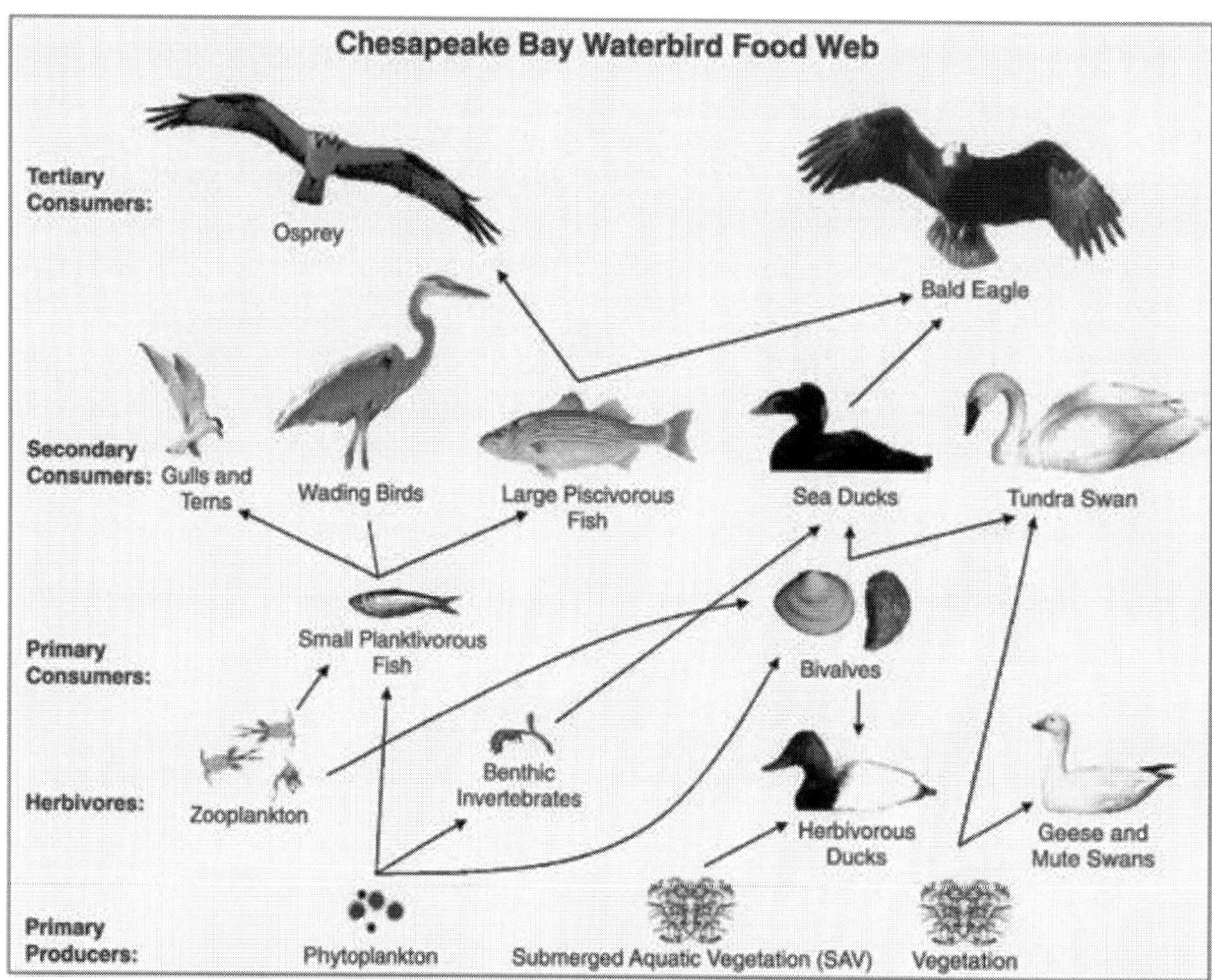

Picture from http://commons.wikimedia.org

Questions:

1) What would happen to the Zooplankton population if the Bivalves were to go extinct?

2) What would happen to the fish populations if wading birds and gulls left the area?

3) What would happen to the populations of land animals if domesticated dogs were left in the area to breed?

4) What would happen to the populations if chemicals were released into the water that killed all the phytoplankton?

5) This community is in a cold coastal marine ecosystem. What would happen to each population if climate change made it warmer and brought more fish to the area?

6) What would happen to the bird populations if the fishing industry were to overfish the area?

   a. How would that affect the zooplankton population?

   b. How would that affect the osprey population?

7) What would happen to the geese and swan populations if Asian carp were introduced and adapted to this community and ate all the plants?

   a. What would that do to the bivalve populations?

Our Little Mountain

Before I was born, my parents used to go on hikes and have picnics on this small mountain that overlooks a pass between two large mountain ranges. It was filled with tall pine trees with a clearing on the top. One day, shortly after I was born, we drove out to their special spot and saw that bulldozers were getting ready to knock down and uproot the trees. My parents were upset to see this. My dad saw a trailer parked close by with a makeshift parking lot. He went inside to find out what was going on. When he came out, he had a piece of paper in his hand and a scary smile on his face. He told my mom that they were selling land plots and were going to be building a winding road and houses on this mountain. My dad had just purchased the 5-acre lot at the top, which was their special picnic spot. We lived in Golden, Colorado, and he thought it could be a great place to put a vacation home if they could ever afford it one day.

A month later, we came back to see how the construction progressed. All the trees and bulldozers were gone, and a gravel road wound back and forth up our little mountain. My parents had packed a picnic that we would eat on our land at the top. When we had our picnic, the wind kept blowing dirt up into the air, and clouds of dust kept getting in our food and eyes; it was not very enjoyable.

Three years later, after my sister was born and able to walk, we went back to look at our mountain again. There was the greenest grass you ever saw up and down the mountain. Two houses were built, one at the bottom and an A-frame part way up. The one at the bottom served as an office for selling the plots on the mountain. We drove up the mountain and had a great time running around in the grass on our 5-acres. A crow flew down and snatched the peanut butter sandwich right out of my sister's hand. Everyone laughed except for her. Later we saw some deer walk out of the woods next to the land cleared on the mountain. I looked out and imagined John Wayne driving thousands of head of cattle through the pass our land overlooked. My dad got a promotion a few weeks later, and we moved to Wisconsin for his new job.

Ten years later, my dad thought it would be fun for us to go on a family vacation, as he saw in a movie. He wanted to wind around America and eventually see our mountain. We noticed lots of bushes and small trees were all over the mountain when we got there. The house at the bottom was abandoned. My dad did some research while we were there and found out we were only one of two that had purchased any plots, and the developer went bankrupt, and the mountain was abandoned. We went to the top of the mountain and picked berries growing on the bushes there. We saw a few rabbits and a small red fox running down the mountain across the dirt road just before leaving. I thought this could be a great place to put a house one day.

Twenty years later, I was now married with three children. My wife and I thought it would be fun to take our family skiing. I told her that my family owned some land on the top of a little mountain not far away from the lodge where we were staying. While on vacation, one afternoon, we took a drive to find the mountain. We found the little dirt road off the highway. As we drove up, we saw many leafy trees filling the mountain. We drove up the mountain until we reached the end of the road at the top. We walked through the forest until we reached a clearing at the top of the mountain, looking over the large valley pass that separated the mountain ranges on either side. I remembered the other times I had been there when I was younger. I wondered why my parents never built a house up there. There were lots of squirrels and birds. Pine tree seedlings seemed to be taking hold and growing around the clearing.

When I had retired, and our oldest was visiting with our grandson, she asked about the land we visited in Colorado. My parents had died a few years before when COVID 19 hit, and I decided to look through the lockbox my father had given me before their passing. I noticed the deed on the land had been amended. The other owner moved off the mountain when the trees blocked their view of the valley below. My father was sold all the land surrounding the mountain. I now owned our little mountain. My publishing business took off, and I had enough money to build a house on the top of that clearing on the mountain. I hopped on a plane, went straight to our mountain, and saw that most of the leafy trees were gone. There were lots of small and medium-sized pine trees covering the mountain. I hired a construction company and built a large two-story cabin with big picture windows looking over both sides of the mountain. Because it was so far away from the nearest town and it was on the top of the mountain, it took a few years to build. My wife and I moved there after our youngest son graduated college. Our house also became the rest of our family's vacation home my parents dreamed about when I was just a baby.

I was 95 years old when I died. My last view of this Earth was the tall pine trees that were just like the ones I saw when I was a baby. The trees filled the whole mountain like they did when my parents went on their hikes and picnics on the top of the mountain. These trees were the bottom of the frame of our bay window in the front of our house. I do not know if I imagined it or not. But I thought I saw a herd of cattle going through the pass in the valley between the two mountain ranges just before I died.

Directions: Read each section of the story and draw a picture of the story's description in each box. When done with the story and your drawings, you should have a good picture of how succession looks.	**Before I was Born**
A Month Later	**Three Years Later**
Ten Years Later	**Twenty Years Later**
When I Retired	**I was 95 Years Old**

Questions:

1) What happened to the biodiversity when the land developers knocked down the forest?

2) During the story, when did the mountain have its greatest diversity? Explain why.

3) What happened to the plant populations when the forest was knocked down?

    a. Were there any homes for the animals there?

    b. What happened to the animal populations?

4) When did the grass populations increase?

5) When the trees started growing, what happened to the population of the grasses?

6) What happened to the kinds of plants during the story?

7) What happened to the kinds of animals during the story?

8) Who was the invasive species in the story, and how did they affect the ecosystem populations when they arrived?

9) From this story, can life find a way to recover from change? Explain your answer.

10) Who do you think will be outlived, humans or the rest of life? Explain.

Primary or Secondary Succession?

Directions:

Go outside, and find five examples of each.

1) Primary Succession takes place where rock is taking in life and eroding it away to make soil.
    a.
    b.
    c.
    d.
    e.

2) Secondary Succession occurs when something interrupts the ecosystem, reestablishing itself while keeping the soil.
    a.
    b.
    c.
    d.
    e.

Virtual Investigations that go with Ecological Changes

ExploreLearning.com

Food Chain

Prairie Ecosystem Gizmo

Forest Ecosystem Gizmo

Rabbit Population by Season Gizmo

Ecosystems STEM Case

Ecosystems Handbook

Coral Reefs 1 – Abiotic Factors

Coral Reefs 2 – Biotic Factors Gizmo

Natural Selection

Evolution: Mutation and Selection

Evolution: Natural and Artificial Selection

Rainfall and Bird Beaks

Evolution STEM Case

Evolution Handbook

Human Evolution – Skull Analysis

Microevolution

Cladograms

Effect of Environment on New Life Form Gizmo

Estimating Population Size Gizmo

Phet.colorado.net

Natural Selection

8th Grade Science TEKS Correlations

Nature of Science Concept Map Science, Grade 8 b 1ABH 2AD 3AB

Focus on the Process Science, Grade 8 b 1ABCDEFG 2AC 3ABC 5ACD

Measurement Lab Science, Grade 8 b 1ABCDEF 2BC 3AB 5AC

Patterns in Pennies Science, Grade 8 b 1ABCDEF 2BCD 3ABC 5AC

Virtual Investigations for Introduction Science, Grade 8 b 1ABEFG 2ABCD 3ABC 4AC

Elements Compounds and Mixtures Concept Map Science, Grade 8 b 6A

Metal or Nonmetal Science, Grade 8 b 1ABCDEF 3AB 5A 6AB

Periodic Table Activity Science, Grade 8 b 1ABEFG 2B 3AB 5AEF 6AB

Elements Compounds and Mixtures Research Science, Grade 8 b 1ABE 3AB 4B 5ACG 6A

Elements Compounds and Mixtures Science, Grade 8 b 1ABCDEF 3AB 5ABC 6A

Separating Mixtures Science, Grade 8 b 1ABCDE 3AB 5ABC 6A

Separating Pigments Science, Grade 8 b 1ABCDEF 3AB 4C 5ABC 6A

50 + 50 Does Not Equal 100 Science, Grade 8 b 1ABCDE 2BC 3AB 5ABCD 6A

Percent Sugar in Bubble Gum Science, Grade 8 b 1ABCDEF 2AB 3AB 4C 5ABC 6A

Virtual Investigations for Elements Compounds and Mixtures Science, Grade 8 b 1ABEFGH 2ABCD 3AB 5ABCDF 6A

Properties of Water Concept Map Science, Grade 8 b 6C

Building a Model of a Water Molecule Science, Grade 8 b 1ABCDG 2AD 3AB 5ACG 6C

Checking Polarity Science, Grade 8 b 1ABCDEG 2AD 3AB 5ABCG 6C

Celery Transport Science, Grade 8 b 1ABCDEH 3AB 5ABDEFG 6C

Transpiration Pull Science, Grade 8 b 1ABCDEF 3AB 5ABDEFG 6C

Seeing a Stoma Science, Grade 8 b 1ABCDE 3AB 5ABCDEFG 6C

How does Rain Form? Science, Grade 8 b 1ABCDE 3AB 5ABDEG 6C

Virtual Investigations for Properties of Water Science, Grade 8 b 1ABEFGH 2ABCD 3ABC 4AC 5ABCDG 6C

Acids and Bases Concept Map Science, Grade 8 b 6D

Which is an Acid and Which is a Base? Science, Grade 8 b 1AB 3AB 5A 6D

A Homemade Indicator Science, Grade 8 b 1ABCDE 3AB 5ABC 6D

Observing Acid Relief Science, Grade 8 b 1ABCDE 3AB 4C 5ABCG 6D

Acid or Base Grape Juice Indicator Science, Grade 8 b 1ABCDEF 2A 3AB 5ABC 6D

Characteristics of Acids and Bases Science, Grade 8 b 1ABCDEF 3AB 5AC 6D

Which will Corrode a Nail? Science, Grade 8 b 1ABCDEFH 3AB 5ABG 6D

Virtual Investigations for Acids and Bases Science, Grade 8 b 1ABEFGH 2ABCD 3AD 4AC 5ABC 6D

Conservation of Mass Concept Map Science, Grade 8 b 6E

Conservation of Mass Equations Science, Grade 8 b 1ABEG 2BC 3AB 4C 5ABCEG 6E

Home Chemistry Science, Grade 8 b 1ABCDE 3AB 5ABCG 6E

Types of Chemical Reactions Science, Grade 8 b 1ABCDE 3AB 5ABCG 6E

Removing Carbon from Sugar Science, Grade 8 b 1ABCDE 2BC 3AB 5BEG 6E

Conservation of Mass Science, Grade 8 b 1ABCDE 2BC 3AB 4B 5ABCDEG 6E

The Law of Conservation of Mass Science, Grade 8 b 1ABCDE 2BC 3AB 5ABCDEG 6E

Conservation of Life: Photosynthesis and Respiration Science, Grade 8 b 1ABG 2BC 3ABC 4C 5ABCDEFG 6E

Virtual Investigations for Conservation of Mass Science, Grade 8 b 1ABEFGH 2ABC 3AB 4AC 5ABCDEFG 6E

Accelerated Force and Motion Concept Map Science, Grade 8 b 7AB

Marbles in Motion Science, Grade 8 b 1ABCDEF 2BC 3AB 5ABCDEG 7AB

Ball Bounce Science, Grade 8 b 1ABCDEF 2B 3AB 5ABCDEG 7A

Cart on a Ramp Science, Grade 8 b 1ABCDEF 2B 3AB 5ABCDEG 7A

Picket Fence Free Fall Science, Grade 8 b 1ABCDEF 2BC 3AB 5ABCDEG 7A

Measuring the Effects of Air Resistance Science, Grade 8 b 1ABCDEF 2BC 3AB 4C 5ABCDEG 7AB

Elevator Lab Science, Grade 8 b 1ABCDE 2B 3AB 5ABCDEG 7AB

Newton's Relay Race Science, Grade 8 b 1ABCDE 3AB 5ABCDEG 7AB

Newton's Second Law Science, Grade 8 b 1ABCDEF 2BC 3AB 5ABCDEG 7AB

Fan Cart Lab Science, Grade 8 b 1ABCDEF 2ABCD 3AB 5ABCDEG 7A

Water Bottle Rockets Science, Grade 8 b 1ABCDE 2B 3AB 5ABCDEG 7AB

Virtual Investigations for Accelerated Force and Motion Science, Grade 8 b 1ABEFGH 2ABCD 3ABC 4AC 5ABCDEG 7AB

Waves Concept Map Science, Grade 8 b 8AB

Measuring Wave Properties Science, Grade 8 b 1ABCDEF 2BC 3AB 5ABCDEG 8A

Observing Waves in a Slinky Science, Grade 8 b 1ABCD 3B 5ABCDEG 8A

Observing Sound Science, Grade 8 b 1ABCDE 3AB 5ABE 8A

Coffee Can Phones Science, Grade 8 b 1ABCDE 3AB 5ABDEG 8A

Music Test Tubes Science, Grade 8 b 1ABCDEFH 2B 3AB 4C 5ABCDEG 8A

Singing Glasses and Dancing Toothpick Science, Grade 8 b 1ABCDE 2B 3AB 5ABDEG 8A

Playing the Rubber Band Science, Grade 8 b 1ABCDEF 2B 3AB 4C 5ABCDEG 8A

Music has Patterns Science, Grade 8 b 1ABCDEF 2BC 3AB 4C 5ABCDE 8A

The Doppler Effect Science, Grade 8 b 1ABCDE 2B 3AB 5ABCDEG 8A

Making a Rainbow Science, Grade 8 b 1ABCDE 2B 3AB 5ABCE 8AB

Polarization of Light Science, Grade 8 b 1ABCDEF 2B 3AB 5ABCDE 8B

3D Glasses Science, Grade 8 b 1ABCDE 2B 3AB 4C 5ABCDE 8B

Light Pipes Science, Grade 8 b 1ABCDE 2B 3AB 4C 5ABDEFG 8B

Water Refraction Science, Grade 8 b 1ABCDEF 3AB 5ABDG 8B

Test Tube Lenses Science, Grade 8 b 1ABCDEF 3AB 5ABCEG 8B

Reflection Lab Science, Grade 8 b 1ABCDEF 2B 3AB 5ABCDEG B8

Magnifying Power Science, Grade 8 b 1ABDCEF 2BC 3AB 5ABCDEG 8B

Brightness and Distance Science, Grade 8 b 1ABCEF 2B 3AB 5ABCDEG 8AB

Uses of the Electromagnetic Spectrum Science, Grade 8 b 1AB 4C 5ACE 8B

How we use Microwave Ovens Science, Grade 8 b 1ABCDEF 2BC 3AB 5ABCDEFG 8B

Virtual Investigations for Waves Science, Grade 8 b 1ABEFGH 2ABCD 3ABC 4AC 5ABCDEFG 8AB

The Development of the Universe Concept Map Science, Grade 8 b 9ABC

Star's Life Cycle Science, Grade 8 b 1ABG 3AB 5ABCDEG 9A

Star Life Cycle Model Science, Grade 8 b 1ABCDEFG 2AD 3AB 5ABCDEG 9A

Our Bright Morning Star: the Sun Science, Grade 8 b 1ABEG 3AB 5BDEG 9A

Nuclear Fission and Fusion in a Star Science, Grade 8 b 1ABEFG 3AB 5ABDEG 9A

Compare and Classify Stars Science, Grade 8 b 1ABEG 2B 3AB 5ACD 9A

Classifying Galaxies Science, Grade 8 b 1ABEG 3AB 5AEG 9B

A Guide to the Milky Way Galaxy Science, Grade 8 b 1ABEG 2BC 3AB 4C 5ABCDEG 9B

Ten Things NASA Wants you to Know... Science, Grade 8 b 1ABEG 2BC 3ABC 5ABCDEG 9C

The History of the Big Bang Theory Science, Grade 8 b 1ABE 3ABC 4AB 5ABCDEG 9C

The Pixel of Our Universe Science, Grade 8 b 1ABCDEGH 3ABC 5ABCDEG 9C

WMAP Science, Grade 8 b 1ABEG 2BC 3ABC 4A 5ABCDEG 9C

Virtual Investigations for the Development of the Universe Science, Grade 8 b 1ABEFGH 2ABCD 3ABC 4AC 5ABCDEG 9ABC

Weather Patterns Concept Map Science, Grade 8 b 10ABC

The Water Cycle Science, Grade 8 b 1ABEFG 2B 3AB 4A 5ABDEG 10A

Seasons and the Tilt of the Earth Science, Grade 8 b 1ABCDEFG 2BC 3ABC 5ABCDEG 10A

How Hurricanes Form Science, Grade 8 b 1ABEG 3AB 5ABCDEG 10ABC

Seeing Patterns in the Layers of the Atmosphere Science, Grade 8 b 1ABEFG 2B 3AB 5ABCDEFG 10A

Global Wind Movement Science, Grade 8 b 1ABEG 3AB 5ABCDEG 10AB

Model Showing the Rotation of the Earth... Science, Grade 8 b 1ABCDEG 2AD 3AB 5ABDEG 10B

A Local Weather Study Science, Grade 8 b 1ABCDEF 2BC 3AB 5ABCEG 10A

Relative Humidity Science, Grade 8 b 1ABCDEF 2AD 3AB 5ABCEG 10A

Temperature Inversions Science, Grade 8 b 1AB 3AB 4B 9C 5ABDEG 10A

Virtual Investigations for Weather Patterns Science, Grade 8 b 1ABEFGH 2ABCD 3ABC 4AC 8ABC 5ABCDEG 10AB

Influences on Climate Change Concept Map Science, Grade 8 b 10AB 11AB

Composition of the Atmosphere Science, Grade 8 b 1AB 3ABC 4B 5ABEFG 10A 11ABC

The Greenhouse Effect Science, Grade 8 b 1ABCDEFGH 2ABCD 3ABC 5ABCDEFG 10A 11AB

Climate and Greenhouse Gases: Data Table Science, Grade 8 b 1ABEF 2B 3ABC 5ABCDEG 10A 11ABC

Carbon Dioxide and Population Science, Grade 8 b 1ABF 2B 3ABC 5ABCDEG 11BC

Carbon Emissions Science, Grade 8 b 1ABEF 2BC 3AB 5ACEG 11BC

Climate Change Science, Grade 8 b 1AB 3ABC 4BC 5ABDEG 11BC

How is Life Allowed on Earth? Science, Grade 8 b 1AB 3ABC 4B 5ABCDEG 10A 11AC

Natural and Manmade Disasters Science, Grade 8 b 1ABEF 3AB 4B 5ABDEG 11ABC

Digital Presentations of Worldwide Disasters Science, Grade 8 b 1ABEF 3AB 4B 5ABDEG 11ABC

Nonrenewable Resources Chart Science, Grade 8 b 1ABEF 3AB 4B 5BDEG 11BC

Renewable Resources Chart Science, Grade 8 b 1ABEF 3AB 4B 5BDEG 11B

Virtual Investigations for Influences of Climate Change Science, Grade 8 b 1ABEFGH 2ABCD 3ABC 4AC 5ABCDEG 10A 11ABC

Structure and Function of Life Units Concept Map Science, Grade 8 b 13AB

Membrane Models Science, Grade 8 b 1ABCDG 2D 3AB 5ABDEFG 13A

Cell Town Science, Grade 8 b 1ABG 3AB 4B 5ABDEF 13A

Characteristics of Prokaryotic and Eukaryotic Cells Science, Grade 8 b 1ABCDEFH 3AB 5ABCDEF 13A

Draw a Detailed Picture of Bacteria Science, Grade 8 b 1AB 3AB 5ACF 13A

Draw a Detailed Picture of Paramecium Science, Grade 8 b 1AB 3AB 5ACF 13A

Paper Mates Science, Grade 8 b 1ABCDEFG 3AB 5ABDEF 12B

Create A Baby Science, Grade 8 b 1ABCDEFG 3AB 5ABDEF 13B

Modeling Cell Division Science, Grade 8 b 1ABCDEG 3AB 5ABCDEFG 13B

Hand Models Showing Cell Division Science, Grade 8 b 1ABC 2AD 3AB 5ACEF 13B

Modeling Meiosis Science, Grade 8 b 1ABCDEG 3AB 5ABCDEFG 13B

Comparing Ratios of Monohybrid Crosses Science, Grade 8 b 1ABCDEFH 2BCD 3AB 5ACG 13B

Making a Karyotype Science, Grade 8 b 1ABDFG 2D 3AB 5ABCFG 13B

Lego Mutation Models Science, Grade 8 b 1ABCD 3AB 5ABCEFG 13BC

Protein Synthesis of a Quaddie Science, Grade 8 b 1ABEFG 2D 3AB 5ABCDEFG 13B

Protein Synthesis Role Play Science, Grade 8 b 1ABG 3AB 5ACDEF 13B

Models of Macromolecules Science, Grade 8 b 1ABCD 2AB 3AB 5ABCDEG 13AB

Virtual investigations for Structures and Functions of Life Units Science, Grade 8 b 1ABEFGH 2ABCD 3ABC 4AC 5ABCDEFG 13AB

Ecological Changes Concept Maps Science, Grade 8 b 13A 14A

Variation Within a Population Science, Grade 8 b 1ABCDEF 2BCD 3ABC 5ACG 13C

Goldfish Evolution Science, Grade 8 b 1ABCDEFH 2BC 3ABC 5ABCDEFG 13C

Changing Environments for Beads Science, Grade 8 b 1ABCDEF 2ABCD 3ABC 5ABCDEFG 13C

Causes of Invasive Species Science, Grade 8 b 1AB 3AB 4B 5BDF 12A

Humans Changing Ecosystems Science, Grade 8 b 1AB 3AB 4B 5BDFG 12A

What Happened to the Food Web? Science, Grade 8 b 1ABG 3AB 5ABCDEFG 12A

Our Little Mountain Science, Grade 8 b 1ABEFG 3ABC 5ABCDEG 12AB

Primary or Secondary Succession? Science, Grade 8 b 1ABCDE 3AB 5ABDG 12B

Virtual Investigations for Ecological Changes Science, Grade 8 b 1ABEFGH 2ABCD 3ABC 4AC 5ABCDEFG 13C 12AB

Equipment List for All Investigations

If you want to be able to do all the labs in this manual, here is a list of all the equipment you will need in order of appearance:

Small Legos sets

Water

Scale

Meter sticks

Temperature probes

Interface

Computers with Logger Pro

100 mL graduated cylinders

Stopwatch

Rulers

Pennies

Roll of pennies

Empty penny rolls

Internet

Textbooks

Aluminum foil

Milk

Laser pointer

Granite countertop samples

Kool-Aid

Salt

Pencils

Chalk

Muddy water

Magnets

Plastic baggies

Wire strainers

Coffee filters

Hotplates

Sand

Sugar

Marbles

Iron filings

Granola

Beakers

Goggles

Scissors

Pens

Markers

Eyedroppers/pipettes

Nail polish remover

Rubbing alcohol

Chromatography paper

Test tubes

Test tube racks

Paper clips

Rubber stoppers

Bubble gum

Round balloons of different sizes

Molecular model kits

Periodic Table

Paper towels

Celery

Food coloring (many colors)

White carnations

Pressure sensor and tube setup

Plant branch

Scotch tape

Lettuce

Slides and coverslips

Compound light microscopes

Glasses

Ice

Aprons

Red cabbage

Shampoo

Grapefruit juice

Sprite

Ammonia

100% purple grape juice

Antacid tablets

Vinegar

Lemon juice

Drain cleaner

Detergent

Baking soda

Litmus blue paper

Litmus red paper

Universal indicator pH paper

pH meter

Bottled water

Sink

Club soda

Clear liquid soap

Tiny cups

Nails

Bottles of Coke

Orange juice

Hydrogen peroxide

Aluminum pans

Digital scales

Steel wool

Matches

Copper sulfate

Screws

Long-neck lighter

Beaker tongs

Lighter fluid

Ceramic bowl

Ziploc bags

Potassium iodide

Erlenmeyer flasks

Lead nitrate solution

Hot wheels track

Stickers

Small wire baskets

Large bouncy balls

Motion detectors

Spring carts

Vernier Dynamics Systems

Picket fences

Photogates

Ring stands and clamps

Paper

Hanging masses

Elevator

Brooms

Bowling balls

Basketball

Kid's rubber balls

Carts

Rubber bands

Dual-range force sensors

Low g accelerometers

Fans for carts

Water bottle rocket launcher

2-liter bottles

Air pumps with pressure gauge

Telephone cords

Slinkies

Wire hangers

String

Coffee cans

Flat toothpicks

Crystal drinking glasses

Plastic tubs

Digital keyboard

Microphone probe

Football that whistles

Water hose/spray bottle

Diffraction gradient glasses

Polarizing filters

Light sensors

3D glasses

Fiber optics

Square tanks

Erlenmeyer flasks

Rubber stoppers

Flat mirrors

Protractors

Light sources

Magnifying glasses

Laser lights

Light sensors

Microwave oven

Oven mitts

Compass drawing tools

Golf balls

Beads of different colors

150-watt incandescent light sources

Globes

Spoons

Glass bowls

Pepper

Humidity probes

UVB sensors

Shoelaces

Black paint

Press'n Seal sealing wrap

Dishwashing liquid

Twisty ties

Shallow dish

Dry beans

Buckets

Pipe cleaners

Colored pencils

Masking tape

Connecting alphabet baby letters or shapes

Leaves

Shelled nuts

Cheese-flavored goldfish crackers

Pretzel-flavored goldfish crackers

Colored construction paper

SEVEN SIDES PUBLISHING

Made in the USA
Columbia, SC
19 June 2023